Application of Signal Processing Tools and Artificial Neural Network in Diagnosis of Power System Faults

Application of Signal Processing Tools and Artificial Neural Network in Diagnosis of Power System Faults

Nabamita Banerjee Roy

and

Kesab Bhattacharya

CRC Press
Taylor & Francis Group
Boca Raton London New York

CRC Press is an imprint of the
Taylor & Francis Group, an **informa** business

MATLAB® is a trademark of The MathWorks, Inc. and is used with permission. The MathWorks does not warrant the accuracy of the text or exercises in this book. This book's use or discussion of MATLAB® software or related products does not constitute endorsement or sponsorship by The MathWorks of a particular pedagogical approach or particular use of the MATLAB® software.

First edition published 2022
by CRC Press
6000 Broken Sound Parkway NW, Suite 300, Boca Raton, FL 33487-2742
and by CRC Press
2 Park Square, Milton Park, Abingdon, Oxon, OX14 4RN

CRC Press is an imprint of Taylor & Francis Group, LLC

© 2022 Nabamita Banerjee Roy and Kesab Bhattacharya

ISBN: 978-0-367-43113-6 (hbk)
ISBN: 978-1-032-04363-0 (pbk)
ISBN: 978-0-367-43114-3 (ebk)

Typeset in Times
by SPi Global, India

Contents

Authors

Nabamita Banerjee Roy is presently Associate Professor in the Electrical Engineering Department of Narula Institute of Technology, Kolkata, India. She has obtained her graduation in Electrical Engineering from B.E. College, Shibpur, Howrah, West Bengal (presently IIEST Shibpur) in 2002. She has obtained both MEE and PhD from Jadavpur university, Kolkata in 2004 and 2017, respectively. She has a rich and diverse academic career as a faculty in Electrical Engineering Degree level and as an administrator (acting principal) at the diploma engineering level, since 2004. She has also served as HOD, Electrical Engineering Department at Narula Institute of Technology. She has supervised many BTech and MTech projects. She has published papers in National/International conferences and journals along with a Book Chapter in Springer. Presently, she is the supervisor of PhD candidate with the title of research proposal as *Identification and Characterisation of Low and High Impedance Faults Using Signal Processing Techniques*. Her areas of research include signal processing, power system faults, neural network, soft computation, and high voltage engineering.

Kesab Bhattacharya earned a BE and an MEE (high-voltage engineering) and a PhD at Jadavpur University (JU), Kolkata, India, in 1982, 1984, and 2000, respectively. He is a Professor of the Department of Electrical Engineering, JU. He worked with NGEF Ltd. for 6 months as a Marketing Engineer and with General Electric Company (India) as a Design Engineer of HT motors from April 1984 to October 1987. He has guided many MEE and PhD students and published several research papers in national and international journals.

Introduction

Transmission-line relaying involves three major tasks, namely detection, classification, and identifying the location of transmission-line faults. Fast detection of transmission-line faults enables quick isolation of the faulty line from service and hence protects it from the harmful effects of the fault. Classification of faults means identification of the type of fault, and this information is required for fault location and accessing the extent of repair work to be carried out. Accurate, fast, and reliable fault classification technique is an important operational requirement in modern-day power transmission systems. On the other hand, the information of the type of fault is needed for fault location estimation. Because of these requirements, a significant amount of research work has been directed to address the problem of an accurate fault classification scheme.

The conventional methods of fault classification involve complex mathematical operations. The complexity of the calculations increases with the increase in size of the power system network. The calculations require the data of line parameters of the system components: the positive, negative, and zero sequence impedances. A fault classification technique for distribution systems is proposed in [1] by the modeling of sequence networks. The results have shown satisfactory accuracy but the speed of the given method has not been mentioned. It is not clear whether the method presented in [1] depends on the parameters, such as fault resistance, fault location, and fault inception angle (that are not accessible).

The soft computing techniques have shown relatively better performance in the method of fault classification with respect to speed and accuracy. The methods mainly involve the simulations of network and faults in reliable softwares like EMTP, PSCAD, and MATLAB®, involving the application of signal processing tools, i.e., Wavelet transform (WT) and S-Transform (ST).

Discrete Wavelet Transform (DWT) is a powerful signal analysis tool which has been extensively used for fault detection in transmission lines. Several distinctive features are mainly extracted from the line current or voltage signals after they are being processed through DWT [2]. Subsequently, the aforesaid extracted features are fed to a Genetic Algorithm-based fault classifier [3] or neural network for fault classification [4–5].

A new percentage differential protection scheme for double-circuit transmission line using WT is presented in [6] but the effect of noise on the current signals and computation of fault location has not been investigated.

A novel wavelet-based methodology has been developed in [7] for real-time fault-induced transient detection in transmission lines. The proposed method in this article is based on computation of wavelet coefficient energy, which takes into account the effect of noise, and is independent of the choice of mother wavelet. However, the issues of fault classification and estimation of fault location are not included in this paper.

The challenging task of protection of multiterminal transmission lines is addressed in [8] using a support vector machine classifier. WT has been used for the decomposition of measured signals into different frequency bands followed by subsequent calculation of signal energy at each frequency band. The normalized value of signal energy has been used as SVM input. The training and testing data have been generated considering various system parameters such as fault inception angle, fault resistance, and fault location. A denoising process has been performed on the signals to increase the noise immunity of the proposed protection algorithm. However, the fault location has not been estimated in this paper.

Fuzzy-neuro approaches have also been proposed for classification in some papers [9–10]. An efficient technique has been proposed in [9] for the classification and localization of transmission line faults under different conditions. The effect of noise on the current signals has not been considered here.

Pattern recognition approach is established in quick and accurate identification of the fault components and fault sections. A novel method for high impedance fault (HIF) detection based on pattern recognition systems is presented in [11] where WT is used for the decomposition of signals and feature extraction. WT-based denoising technique was employed before the signal features were extracted by a Bayes Classifier. A detailed analysis of HIF detection using real-time data is presented in [12] in which WT has been used to extract the high-frequency content of the phase voltage signals. The proposed algorithm does not include the effect of noise on the practical data. WT-based denoising method has been also employed in [13] after which the denoised signals have been decomposed by DWT for feature extraction. The extracted features have been used as ANN inputs for HIF identification. Fault location estimation has not been investigated in [11–13].

It is well established in [14–15] that wavelet energy entropy can be used as an important feature in classifying faults in transmission lines. A fault detection and classification algorithm is successfully demonstrated in [14] with wavelet entropy principle. In this paper, the current waveforms of all the phases are decomposed using DWT under different fault conditions. The wavelet entropies of the decomposed signals are used as features for fault analysis. An expert system based on wavelet entropy and artificial neural network is demonstrated in [15] for fault classification and distance estimation in an overhead transmission line. The training data set is only generated for different fault locations without considering variation of fault resistance and fault inception angle. The effect of noise on the current/voltage signals has not been considered.

In the DWT-based technique, an appropriate mother wavelet and the number of the decomposition level must be chosen by trial-and-error procedure. On the other hand, Continuous Wavelet Transform (CWT) gives much more detailed information of a signal with higher computational burden. The performance of WT is significantly degraded in real practice under noisy environment. On the other hand, ST has the ability to detect the disturbance correctly in the presence of noise due to which it is very popular in detecting power system faults and disturbances. ST is a modified version of CWT which retains the absolute phase of every frequency component. In [16–19], several power quality events and disturbances have been thoroughly investigated and diagnosed by ST in conjunction with neural or fuzzy network.

A radial basis function neural network (RBFNN) has been proposed in [20] for fault classification and location estimation after pre-processing the current and voltage signals using Hyperbolic ST. A new approach of transmission line protection has been demonstrated in [21] in which the current and voltage signals are processed by STs. The change in spectral energy of the ST of the current and voltage signals provides the information regarding fault detection. The fault location is calculated using the polynomial curve fitting technique. The fault classification is based on the threshold value of the signal energy and no expert system consisting of ANN or SVM has been developed. A spectral energy function-based fault detection is presented in [22] during a power swing using a ST. The proposed technique is thoroughly tested for different fault conditions during a power swing with possible variations in operating parameters, but there is no scope for estimation of fault location in the given method.

This book presents a ST-based PNN classifier for fault classification where some features are required for detecting a type of fault and the affected phase. The voltage and/or current signals of the three phases are processed through ST to generate complex S-matrices. The features extracted from ST are given to PNN for training, and subsequently, it is tested for an effective classification. After detecting the affected phase, the major harmonic component of the voltage signal of the faulty phase is used for training the BPNN for obtaining the fault location. All the power system networks involved in the study have been simulated in MATLAB Simulink environment. The feature extraction is done by programming in MATLAB. Since a maximum of six features are required for fault detection and one feature for estimating fault location, the memory requirement and computation time will significantly reduce. Moreover, using ST instead of WT will avoid the requirement of testing various families of wavelets in order to identify the best one for detection.

Decision-tree (DT)-based classifier has shown promising results as a classifier. [23] presents a HIF detection method based on DTs. A scheme of fault classification in single-circuit and double-circuit transmission lines is suggested in [24–25]. In all these papers, the classifier is not tested under noisy environment. DT-based classifier has once again shown excellent performance in [26] where a new approach for fault zone identification and fault classification for thyristor controlled series compensator (TCSC) and unified power flow controller (UPFC) line using DT is presented.

The PNN can function as a classifier [27–30] and has the advantage of being a fast-learning process, as it requires only a single-pass network training stage without any iteration for adjusting weights. Further, it can adapt itself to architectural changes. As the structure of the PNN is simple and learning efficiency is very fast, it is suitable for signal detection problems. It has been established in [28] that PNN-based method is superior in distinguishing the fault transients than the Hidden Markov Model (HMM) and DT classifier with higher accuracy. The suggested PNN classifier is also tested with simulated voltage signals contaminated with synthetic noise. The ability of PNN as a fast and precise fault classifier is also established in [29] where it is compared with the feed-forward neural network and the radial basis function network. A new combined method of ST and PNN is demonstrated in [30] for differential protection of power transformers.

A novel technique has been established in [31] for determining fault location in parallel transmission lines where wavelet analysis is used for fault detection and classification. The problem of locating single line-to-ground faults has been examined in [32] for multiring electrical distribution networks. The developed method is based using artificial neural network techniques as the algorithm to save time in fault location, as well as estimating the value of the fault resistance. A simple and accurate fault location algorithm is developed in [33] that is feasible for all types of faults without involving line parameters. However, the identification of the faulty phase in cases of LG, LL, and LLG types of faults have not been discussed here.

Phasor measurement unit-based fault location techniques are proposed in [34]. This article regenerates the method using only voltage measurements and suggests a new algorithm using both voltage and current measurements with a higher degree of accuracy. The proposed technique is also tested with superimposed noisy test data of signal to noise ratio (SNR) of 20 dB. But no separate algorithm is mentioned for fault classification. Pattern classification technology is developed in [35] for determination of fault location based on PMU measurements. Support Vector Regression (SVR) is established to be an efficient tool in [36,37] for estimation of fault location. A novel transmission line fault location algorithm is demonstrated in [36] based on wavelet packet transform (WPT) combined with SVR without any classification scheme being proposed to determine the type of fault. A combination of Stationary Wavelet Transform (SWT), SVM, Determinant Function Feature (DFF), and SVR is employed to develop a fault classifier and a fault locator in [37]. SVM is used for forming DFF. SWT combined with DFF performs the feature extraction for fault classification. An algorithm has been is developed combining SWT, DFF, and SVR to locate the fault distance.

A differential equation-based distance protection algorithm is proposed in [38] to locate the phase-ground type of faults. Fault classification is not addressed in this paper. A novel technique of detecting only the different types of phase(s) – ground faults is presented in [39]. The proposed method does not distinguish between phase–phase and three-phase short-circuited faults. An entirely new method called Morphology Singular Entropy (MSE) is presented in [40] for selecting the faulty phase of transmission line. The suggested technique is insensitive to noise.

The scope of the present book is limited to the application of standard ST using Gaussian window. During handling of large size data, the method may be unsuitable due to high computational burden of standard ST. To improve its computational efficiency, the discrete orthonormal Stockwell transform (DOST) is proposed in [41]. A computationally fast version of discrete ST is presented in [42] where cross-differential protection scheme for power transmission systems is proposed.

Another disadvantage of standard ST is its poor time resolution during the onset of a transient event. It suffers from poor energy concentration in the time–frequency domain. A fast adaptive discrete generalised ST (FDGST) algorithm is presented in [43] in which the algorithm optimizes the shape of the window function for each analysis frequency to improve the energy concentration of the time–frequency distribution. Hyberbolic S-Transform (HST) has shown promising results in [44] where a method is proposed to discriminate most important transient fault currents of transformers in which ST uses a hyperbolic window function. A new full-scheme distance protection for a series-compensated transmission line is proposed in [45]. The new combination of hyperbolic HST and learning machines is applied for fault detection, classification, and location, which are the three main aspects of distance relays. The HST is used for extracting useful features from the current and voltage signal sample of the power system from one terminal. The extracted features are employed for distance protection using support vector classification and SVR methods. A novel technique of determination of fault location has been demonstrated in [46] in which the method requires the synchronized measurement of the current signals at all terminals. ST has been employed on the modal transformation of the current signals. The rows of the ST matrix have been investigated and the best row has been selected to give the accurate fault location. A criterion for selecting the best row has been developed. There is no application of neural network. An elaborate algorithm has been developed for identification of the faulty phase, faulty section, and fault-location estimation. The results have been shown for a three-terminal and a six-terminal network. The detection of ground fault has not been mentioned. The effect of noise on the current signals has not been discussed. Four new methods have been introduced and compared in [47] for detection of fault zone in series compensated lines. The fault classification and determination of the exact fault location have not been discussed here. PMU-based fault identification in a power grid has been presented in [48] in which FFT has been used for phasor estimation and SVM as a classifier. A novel algorithm based on DWT has been developed in [49] for fault classification in high voltage transmission lines. [50] represents a Finite State Machine (FSM)-based method of fault identification and determination of the faulty phase in a 400 kV double-circuit transmission system. The main principle behind the working of an FSM-based detector and classifier is extrapolating the changes occurring in fundamental current signals from pre-fault condition to post-fault condition. The proposed method in [50] involves DFT for estimating the fundamental values of current at discrete intervals of time.

The paper in [51] presents a non-unit protection scheme for series capacitor compensated transmission lines (SCCTL) using DWT and k-nearest neighbor (k-NN) algorithm. The signal processing and feature extraction have been done using DWT due to its capability to differentiate between high and low frequency transient components. The approximate wavelet coefficients of level 1 have been considered for feature extraction to classify the types of faults, while those of level 3 have been considered for obtaining fault location by k-NN algorithm. The proposed method has been tested for all possible fault conditions with 100% accuracy achieved in fault classification and with $\leq 1\%$ error in obtaining fault location. A method of fault classification and determining its location has been proposed in [52] based on Fast Discrete S-Transform and the viability of the method has been tested by placing STATCOM in the middle of one of the transmission lines. A decision tree regression (DTR)-based fault distance estimation scheme for double-circuit transmission lines is presented in [53]. Three-phase current and voltage signals at one end of a network have been processed by both DFT and DWT. A comparative study of both the techniques has been carried out to observe the effect of signal processing on the fault location estimation method. The proposed method is tested on 2-bus, WSCC-9-bus, and IEEE 14-bus test systems. The test results confirm that the proposed DTR-based algorithm is not affected by the variation in fault type, fault location, fault inception angle, fault resistance, pre-fault load angle, SCC, load variation, and line parameters. The proposed scheme is relatively simple and easy in comparison with complex equation-based fault location estimation methods. A novel approach of fault detection, classification, and localization has been provided in [54] where DWT has been for feature extraction from voltage and current signals. The features have been trained by SVM to obtain the desired output. A scheme of fault detection and classification in AC transmission lines has been presented in [55] where DWT has been once again used for decomposition of current signals at different levels to form a wavelet-covariance matrix and from this matrix power spectral density (PSD)is calculated. PSD is the key feature for fault detection and the classification is accomplished via the Hellinger distance. Another method of fault detection and classification has been suggested in [56] with respect to STATCOM placed in the middle of the transmission line and DWT has been used to obtain Spectral Energy content of current signal of each phase at both sending and receiving ends of the network. Different Spectral Energy content of each phase current is calculated which forms the key feature of fault recognition. The paper in [57] demonstrates a new digital relaying for detection, classification, and localization of faults on the hybrid transmission line consisting of an overhead line and an underground cable. Here Fast Discrete Orthogonal S-Transform (FDOST) has been implemented for processing of three-phase fault current signals at one end of the line and entropy of FDOST coefficients is extracted as the feature. Both SVM classifier model and SVR model are employed for pattern recognitions to predict the types and locations of faults. A fundamental phasor-based approach has been provided in [58] for development of a scheme of fault detection, classification, and location for two-terminal long transmission lines. DFT has been implemented for obtaining current phasor values and the proposed method has been tested under various fault conditions. A novel communication-based HIF protection scheme is proposed in [59] on the basis of WPT and extreme learning machine (ELM). WPT is utilized to extract high-frequency coefficients of three-phase currents of both ends of transmission line and ELM identifies the faulty phase(s). Another algorithm of fault identification has been proposed in [60] where WPT is used for calculation of energy coefficients of the current signals. The proposed method is insensitive to variation in load and does not need retesting even if the transmission system configurations are changed.

A method of fault classification on overhead transmission line and determination of fault location has been presented in this book. The proposed technique is based on multi resolution ST and artificial neural network. Satisfactory results have been obtained by implementing the technique on a single-circuit a.c. transmission system under balanced and unbalanced loading conditions. The method has been further tested on a multiterminal a.c. system and a six-pulse HVDC transmission system. In both the cases the results have been convincing.

The rest of the book is organized as follows. Chapter 1 presents a brief overview of power system faults. A compressive study of WT has been provided in Chapter 2 with case studies and examples. Chapter 3 presents a description on ST. Application of ST on different electrical signals has been discussed in Chapter 4. A generalized description of neural network and the mathematical formulations have been provided in Chapter 5. A method of fault identification and determination of fault location has been demonstrated on a single-circuit a.c. transmission system in Chapter 6. Signal-energy-based fault classification has been implemented on a single-circuit a.c.

transmission system under unbalanced loading condition in Chapter 7. The proposed method is further investigated on a multiterminal system in Chapter 8. A combined method of ST and ANN has been implemented for fault classification and determination of fault location in a HVDC system in Chapter 9. The conclusion of the entire book is summarized in Chapter 9.

REFERENCES

1. M. Abdel-Akher and K. Mohamed Nor, "Fault analysis of multiphase distribution systems using symmetrical components", *IEEE Transactions on Power Delivery*, vol. 25, no. 4, pp. 2931–2939, Oct 2010.
2. D. Chanda, N. K. Kishore, and A.K. Sinha, "Application of wavelet multiresolution analysis for identification and classification of faults on transmission lines", *Electric Power Systems Research*, vol. 73, pp. 323–333, 2005.
3. J. Upendar, C. P. Gupta, and G. K. Singh, "Discrete wavelet transform and genetic algorithm based fault classification of transmission systems", *15th National Power Systems Conference*, IIT Bombay, Dec 2008, pp. 323–328.
4. P.S. Bhowmik, P. Purkait, and K. Bhattacharya, "A novel wavelet transform aided neural network based transmission line fault analysis method", *International Journal of Electrical Power and Energy Systems*, vol. 31, no. 5, pp. 213–219, June 2009.
5. Ngaopitakkul S. Bunjongjit, "An application of a discrete wavelet transform and a backpropagation neural network algorithm for fault diagnosis on single-circuit transmission line", *International Journal of Systems Science*, 2012. doi: 10.1080/00207721.2012.670290.
6. Nuwan Perera, and Athula D. Rajapakse, "Series-compensated double-circuit transmission-line protection using directions of current transients", *IEEE Transactions on Power Delivery*, vol. 28, no. 3, pp. 1566–1575, July 2013.
7. Flavio B. Costa, "Fault-induced transient detection based on real-time analysis of the wavelet coefficient energy", *IEEE Transactions on Power Delivery*, vol. 29, no. 1, pp. 140–153, Feb 2014.
8. Flavio B. Costa, "High-frequency transients-based protection of multiterminal transmission lines using the SVM technique", *IEEE Transactions on Power Delivery*, vol. 28, no. 1, pp. 188–196, Jan 2013.
9. M. Jayabharata Reddy and D.K. Mohanta, "A wavelet-fuzzy combined approach for classification and location of transmission line faults", *Electrical Power and Energy Systems*, vol. 29, pp. 669–678, 2007.
10. Thai Nguyen and Yuan Liao, "Transmission line fault type classification based on novel features and neuro-fuzzy system", *Electric Power Components and Systems*, vol. 38, no. 6, pp. 695–709.
11. Ali-Reza Sedighi, Mahmood-Reza Haghifam, O. P. Malik, and Mohammad-Hassan Ghassemian, "High impedance fault detection based on wavelet transform and statistical pattern recognition", *IEEE Transactions on Power Delivery*, vol. 20, no. 4, pp. 2414–2421, Oct 2005.
12. Doaa khalil Ibrahim, El Sayed Tag Eldin, Essam M. Aboul-Zahab, and Saber Mohamed Saleh, "Real time evaluation of DWT-based high impedance fault detection in EHV transmission", *Electric Power Systems Research*, vol. 80, 2010, pp. 907–914.
13. Ibrahem Baqui, Inmaculada Zamora, Javier Mazón, and Garikoitz Buigues, "High impedance fault detection methodology using wavelet transform and artificial neural networks", *Electric Power Systems Research*, vol. 81, pp. 1325–1333, 2011.

14. S. El Safty and A. El Zonkoly, "Applying wavelet entropy principle in fault classification", *Electrical Power Energy Systems*, vol. 31, no. 10, pp. 604–607, 2009.

15. Aritra Dasgupta, Sudipta Nath, and Arabinda Das, "Transmission line fault classification and location using wavelet entropy and neural network", *Electric Power Components and Systems*, vol. 40, no. 15, pp. 1676–1689, 2012.

16. P. K. Dash, B. K. Panigrahi, and G. Panda, "Power quality analysis using S-transform", *IEEE Transactions on Power Delivery*, vol. 18, no. 2, pp. 406–411, Apr 2003.

17. M. V. Chilukuri and P. K. Dash, "Multiresolution S-transform-based fuzzy recognition system for power quality events", *IEEE Transactions on Power Delivery*, vol. 19, no. 1, pp. 323–330, Jan 2004.

18. F. Zhao and R. Yang, "Power-quality disturbance recognition using S-transform", *IEEE Transactions on Power Delivery*, vol. 22, no. 2, pp. 944–950, Apr 2007.

19. S. Mishra, C. N. Bhende, and B. K. Panigrahi, "Detection and classification of power quality disturbances using S-transform and probabilistic neural network", *IEEE Transactions on Power Delivery*, vol. 23, no. 1, pp. 280–287, Jan 2008.

20. S. R. Samantaray, P. K. Dash, and G. Panda, "Fault classification and location using HS-transform and radial basis function neural network", *Electric Power Systems Research*, vol. 76, pp. 897–905, 2006.

21. S. R. Samantaray and P. K. Dash, "Transmission line distance relaying using a variable window short-time Fourier transform", *Electric Power Systems Research*, vol. 78, pp. 595–604, 2008.

22. S. R. Samantaray, R. K. Dubey, and B. Chitti Babu, "A novel time–frequency transform based spectral energy function for fault detection during power swing", *Electric Power Components and Systems*, vol. 40, no. 8, pp. 881–897, 2012.

23. Y. Sheng and S. M. Rovnyak, "Decision tree-based methodology for high impedance fault detection", *IEEE Transactions on Power Delivery*, vol. 19, no. 2, pp. 533–536, Apr 2004.

24. A. Jamehbozorg and S. M. Shahrtash, "A decision-tree-based method for fault classification in single-circuit transmission lines", *IEEE Transactions on Power Delivery*, vol. 25, no. 4, pp. 2190–2195, Oct 2010.

25. A. Jamehbozorg and S. M. Shahrtash, "A decision-tree-based method for fault classification in double-circuit transmission lines", *IEEE Transactions on Power Delivery*, vol. 25, no. 4, pp. 2184–2188, Oct 2010.

26. S. R. Samantaray, "Decision tree-based fault zone identification and fault classification in flexible AC transmissions-based transmission line", *IET Generation, Transmission & Distribution*, vol. 3, no. 5, p. 425–436, 2009.

27. M. Tripathy, R. P. Maheshvari, and H. K. Verma, "Probabilistic neural-network-based protection of power transformer", *IET Electric Power Applications*, vol. 1, no. 5, pp. 793–798, 2007.

28. N. Perera, and A. D. Rajapakse, "Recognition of fault transients using a probabilistic neural-network classifier", *IEEE Transactions on Power Delivery*, vol. 26, no. 1, pp. 410–419, Jan 2011.

29. Maryam Mirzaei, Mohd Zainal Abidin Ab. Kadir, Hashim Hizam, and Ehsan Moazami, "Comparative analysis of probabilistic neural network, radial basis function, and feed-forward neural network for fault classification in power distribution systems", *Electric Power Components and Systems*, vol. 39, no. 16, pp. 1858–1871, 2011.

30. Hosung Jung, Young Park, Moonseob Han, Changmu Lee, Hyunjune Park, and Myongchul Shin, "Novel technique for fault location estimation on parallel transmission lines using wavelet", *Electrical Power and Energy Systems*, vol. 29, pp. 76–82, 2007.

31. Z. Moravej, A. A. Abdoos, and M. Sanaye-Pasand, "A new approach based on S-transform for discrimination and classification of inrush current from internal fault currents using probabilistic neural network", *Electric Power Components and Systems*, vol. 38, no. 10, pp. 1194–1210, 2010.

32. Meshal Al-Shaher, Ahmad S. Saleh, and Manar M. Sabry, "Estimation of fault location and fault resistance for single line-to-ground faults in multi-ring distribution network using artificial neural network", *Electric Power Components and Systems*, vol. 37, no. 7, pp. 697–713, 2009.

33. Wanjing Xiu and Yuan Liao, "Accurate transmission line fault location considering shunt capacitances without utilizing line parameters", *Electric Power Components and Systems*, vol. 39, no. 16, pp. 1783–1794, 2011.

34. S. F. Mekhamer, A. Y. Abdelaziz, M. Ezzat and T. S. Abdel-Salam, "Fault location in long transmission lines using synchronized phasor measurements from both ends", *Electric Power Components and Systems*, vol. 40, no. 7, pp. 759–776.

35. Ya-Gang Zhang, Zeng-Ping Wang, Jin-Fang Zhang, and Jing Ma, "Fault localization in electrical power systems: A pattern recognition approach", *Electrical Power and Energy Systems*, vol. 33, pp. 791–798, 2011.

36. A. A. Yusuffa, C. Feia, A. A. Jimoha, and J. L. Munda, "Fault location in a series compensated transmission line based on wavelet packet decomposition and support vector regression", *Electric Power Systems Research*, vol. 81, pp. 1258–1265, 2011.

37. A. A. Yusuff, A. A. Jimoh, and J. L. Munda, "Fault location in transmission lines based on stationary wavelet transform, determinant function feature and support vector regression", *Electric Power Systems Research*, vol. 110, pp. 73–83, 2014.

38. Y. Zhong, X. Kang, Z. Jiao, Z. Wang, and J. Suonan, "A novel distance protection algorithm for the phase-ground fault", *IEEE Transactions on Power Delivery*, vol. 29, no. 4, pp. 1718–1725, Aug 2014.

39. Shaofeng Huang, Lan Luo, and Kai Cao, "A novel method of ground fault phase selection in weak-infeed side", *IEEE Transactions on Power Delivery*, vol. 24, no. 5, pp. 2215–2222, Oct 2014.

40. L. L. Zhang, M. S. Li, T.Y. Ji, Q. H. Wu, L. Jiang, and J. P. Zhan, "Morphology singular entropy-based phase selector using short data window for transmission lines", *IEEE Transactions on Power Delivery*, vol. 29, no. 5, pp. 2162–2171, Oct 2014.

41. Y. Wang and J. Orchard, "Fast discrete orthonormal stockwell transform", *SIAM Journal on Scientific Computing*, vol. 31, no. 5, pp. 4000–4012, 2009.

42. K. R. Krishnanand and P. K. Dash, "A new real-time fast discrete s-transform for cross-differential protection of shunt-compensated power systems", *IEEE Transactions on Power Delivery*, vol. 28, no. 1, pp. 402–410, Jan 2013.

43. M. Biswal, and P.K. Dash, "Estimation of time-varying power quality indices with an adaptive window-based fast generalised S-transform", *IET Science, Measurement & Technology*, vol. 6, no. 4, pp. 189–197, 2012.

44. A. Ashrafian, M. Rostami, and G. B. Gharehpetian, "Hyperbolic S-transform-based method for classification of external faults, incipient faults, inrush currents and internal faults in power transformers", *IET Generation, Transmission & Distribution*, vol. 6, no. 10, pp. 940–950, Oct 2012.

45. Z. Moravej, M. Khederzadeh, and M. Pazoki, "New combined method for fault detection, classification, and location in series-compensated transmission line", *Electric Power Components and Systems*, vol. 40, no. 9, pp. 1050–1071, 2012.

46. Alireza Ahmadimanesh and S. Mohammad Shahrtash, "Transient-based fault-location method for multiterminal lines employing S-transform", *IEEE Transactions on Power Delivery*, vol. 28, no. 3, pp. 1373–1380, July 2013.

47. A.A. Razavi, and H. Samet, "Algorithms for fault zone detection in series- compensated transmission lines", *Generation, Transmission & Distribution, IET*, vol. 9, no. 4, pp. 386–394, 2015. doi: 10.1049/iet- gtd.2014.0565.

48. P. Gopakumar, M. J. B. Reddy, and D. K. Mohanta, "Adaptive fault identification and classification methodology for smartpower grids using synchronous phasor angle measurements", *Generation, Transmission & Distribution, IET*, vol. 9, no. 2, pp. 133–145, 2015. doi: 10.1049/iet-gtd.2014.0024.

49. D. Guillen, M. R. Arrieta Paternina, A. Zamora, J. M. Ramirez, and G. Idarraga, "Detection and classification of faults in transmission lines using the maximum wavelet singular value and Euclidean norm", *Generation, Transmission & Distribution, IET*, vol. 9, no. 15, pp. 2294–2302, 2015. doi: 10.1049/iet-gtd.2014.1064.

50. Anamika Yadav, and Aleena Swetapadma, "A finite-state machine based approach for fault detection and classification in transmission lines", *Electric Power Components and Systems*, vol. 44, no. 1, pp. 43–59, 2016. doi: 10.1080/15325008.2015.1091862.

51. P. Swetapadma, Mishra A. Yadav, and A. Y. Abdelaziz, "A non-unit protection scheme for double circuit series capacitor compensated transmission lines," *Electric Power Systems Research*, vol. 148, 2017. doi: 10.1016/j.epsr.2017.04.002.

52. O. Dharmapandit and R. K. Patnaik, "Time frequency response-based current differential protection of wide area compensated power system networks," *International Journal of Power and Energy Conversion*, vol. 8, no. 4, 2017. doi: 10.1504/IJPEC.2017.087322.

53. A. Swetapadma and A. Yadav, "A novel decision tree regression-based fault distance estimation scheme for transmission lines," *IEEE Transactions on Power Delivery*, vol. 32, no. 1, 2017. doi: 10.1109/TPWRD.2016.2598553.

54. N. S. Wani and R. P. Singh, "A novel approach for the detection, classification and localization of transmission lines faults using wavelet transform and Support Vector Machines classifier," *International Journal of Engineering and Technology(UAE)*, vol. 7, no. 2, 2018. doi: 10.14419/ijet.v7i2.17.11559.

55. D. Guillen et al., "Fault detection and classification in transmission lines based on a PSD index," *IET Generation, Transmission and Distribution*, vol. 12, no. 18, 2018. doi: 10.1049/iet-gtd.2018.5062.

56. S. K. Mishra, L. N. Tripathy, and S. C. Swain, "A discrete wavelet transform based STATCOM compensated transmission line," *International Journal of Engineering and Technology(UAE)*, vol. 7, no. 4, 2018. doi: 10.14419/ijet.v7i4.5.20010.

57. B. Patel, "A new FDOST entropy based intelligent digital relaying for detection, classification and localization of faults on the hybrid transmission line," *Electric Power Systems Research*, vol. 157, 2018. doi: 10.1016/j.epsr.2017.12.002.

58. K. Andanapalli, N. Shaik, S. Vudumudi, and B. C. Yenugu, "Fault detection, classification and location on transmission lines using fundamental phasor based approach," *International Journal of Recent Technology and Engineering*, vol. 8, no. 1, 2019.

59. S. AsghariGovar, P. Pourghasem, and H. Seyedi, "High impedance fault protection scheme for smart grids based on WPT and ELM considering evolving and cross-country faults," *International Journal of Electrical Power and Energy Systems*, vol. 107, 2019. doi: 10.1016/j.ijepes.2018.12.019.

60. R. Adly, R. El Sehiemy, M. A. Elsadd, and A. Y. Abdelaziz, "A novel wavelet packet transform based fault identification procedures in HV transmission line based on current signals," *International Journal of Applied Power Engineering (IJAPE)*, 2019. doi: 10.11591/ijape.v8.i1.pp11-21.

1 Power System Faults

1.1 INTRODUCTION

A fault in any electrical system is defined as an abnormal condition during which the current is deviated from its normal path. Fault is inevitable in every power system network and hence development of a suitable protection system is essential. Faults can be categorised into two broad categories as follows:

1. Low Impedance Faults: series and shunt
2. High Impedance Faults: series and shunt

Again, the faults can be of two types: symmetrical and unsymmetrical. Symmetrical faults are those in which the magnitude of current and voltage of each phase remain equal to each other. Asymmetrical faults are those in which both the quantities would be different in each phase. It is required to make a pre-analysis of the possible fault conditions of a network in order to design a suitable relaying system for its protection.

Extensive details of power system faults are available in some important books mentioned in the references. The book in [1] covers an elaborate explanation of modelling and analysis of all types of short-circuit faults in different types of systems. On the other hand, the books from [2–4] contain description of power system faults with various numerical examples.

A Brief analysis of different kinds of faults is discussed in the next section.

1.2 TYPES OF FAULTS AND THEIR ANALYSIS USING CONVENTIONAL TECHNIQUES

Symmetrical faults can be only of single type: i.e., short circuit of the three phases of a power system network. They are all shunt faults. In this case, the magnitude of current and voltage of each phase are the same. Analysis of this type of fault is therefore simple as a three-phase circuit can be reduced to a Thevenin's equivalent circuit across the fault location. The fault current and voltage are calculated for a single phase. The corresponding three-phase magnitudes can be obtained by multiplying the single-phase values by three. An example is shown below.

Figure 1.1 shows a generating station feeding power to a 132 kV system. Calculate the total fault current and fault level. The line is 200 km long.

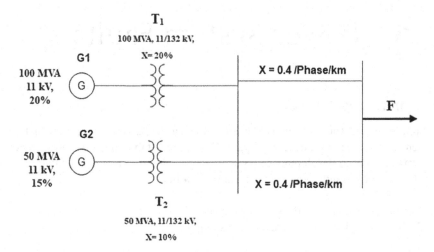

FIGURE 1.1   Single Line Diagram of the system.

Solution: The base MVA for the complete system is 100 MVA, base kV is 11 kV for generator side, and 132 kV for the feeder side.

$$\text{Per unit reactance of generator } G_1 = j0.20 \text{ p.u.}$$

$$\text{Per unit reactance of generator } G_2 = j0.15 \times \frac{100}{50} = j0.30 \text{ p.u.}$$

$$\text{Per unit reactance of transformer } T_1 = j0.20 \text{ p.u.}$$

$$\text{Per unit reactance of transformer } T_2 = j0.1 \times \frac{100}{50} = j0.20 \text{ p.u.}$$

$$\text{Per unit reactance of each line} = j0.4 \times 200 \times \frac{MVA}{(KV)^2}$$

$$= j80 \times \frac{100}{(132)^2} = j0.46 \text{ p.u.}$$

The single line reactance diagram of Figure 1.1 is shown in Figure 1.2, which after further reduction is shown in Figure 1.3 and Figure 1.4, respectively.

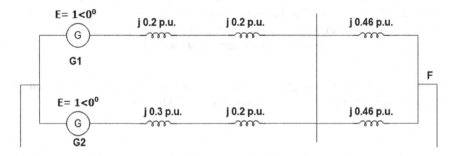

FIGURE 1.2   Reactance Diagram.

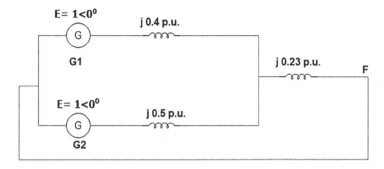

FIGURE 1.3 Equivalent Reactance Diagram of Figure 1.2 after simplification.

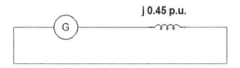

FIGURE 1.4 Equivalent Reactance Diagram of Figure 1.3 after further simplification.

Thevenin equivalent reactance of the system viewed from the fault point F is

$$X_{eqpu} = \frac{j0.4 \times j0.5}{j0.4 + j0.5} + j0.23 = j0.45 \, \text{p.u.}$$

Hence, Fault level $= \dfrac{\text{Base MVA}}{X_{eqpu}} = \dfrac{100}{0.45} = 222 \, \text{MVA}$ **Ans.**

$$\text{Total fault current} = \frac{\text{Fault level in MVA} \times 1000}{\sqrt{3} \times \text{base kV of feeder side}}$$

$$= \frac{222 \times 1000}{\sqrt{3} \times 132} = 971.03 < -90°\text{A, Ans.}$$

Asymmetrical faults can be of both series and shunt type. The series faults are of open circuit type: single phase open and two phase open. Shunt type asymmetrical faults are as follows: Single Line-Ground Fault, Double Line Fault, and Double Line-Ground Fault. Analysis of these faults is different. C.L. Fortesque has provided a theorem for analysis of asymmetrical faults. The details are available in any standard book of Power System [1–4]. The method of analysis can be summarised in the following steps:

1. Formation of single-line diagram (SLD) of the power system network.
2. Formation of positive, negative, and zero sequence network from the SLD.
3. Solving each network to calculate the magnitude of fault current.

An example of a numerical problem is shown below.

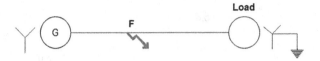

FIGURE 1.5   Single line Diagram.

A three-phase star-connected generator supplies a star-connected inductive load through a transmission line as shown in Figure 1.5. The star point of the load is grounded, and the generator neutral is ungrounded. The load reactance is $j0.5$ p.u. per phase, and the line reactance is $j0.2$ p.u. per phase. The positive, negative, and zero sequence reactances are $j0.6$, $j0.6$, and $j0.06$ p.u., respectively. A single line-to-ground fault takes place in phase 'A' halfway down the line. Prior to fault the network is balanced and the voltage at the fault location is $1 < 0°$ p.u. Calculate the current through the fault path.

Solution: The sequence networks are shown in Figures 1.6–1.8, and the corresponding equivalent sequence impedance of each network is calculated as follows:
Equivalent positive sequence impedance from Figure 1.6 is,

$$Z_1 = \frac{j(0.6+0.1) \times j(0.5+0.1)}{j(0.6+0.1) + j(0.5+0.1)} = j0.323 \text{ p.u.}$$

Equivalent negative sequence impedance from Figure 1.7 is,

$$Z_2 = Z_1 = j0.323 \text{ p.u.}$$

Equivalent zero sequence impedance from Figure 1.8 is,

$$Z_0 = \frac{j(0.06+0.1) \times j(0.5+0.1)}{j(0.06+0.1) + j(0.5+0.1)} = j0.126 \text{ p.u.}$$

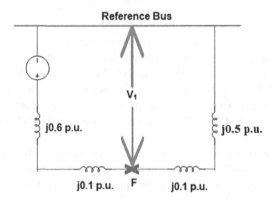

FIGURE 1.6   Positive sequence network.

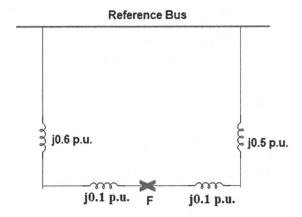

FIGURE 1.7   Negative sequence network.

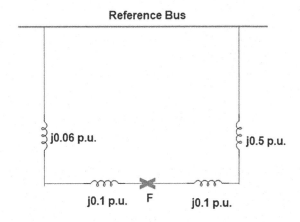

FIGURE 1.8   Zero sequence network.

Fault current,

$$Z_f = \frac{3V_f}{Z_1 + Z_2 + Z_3}$$

$$= \frac{3 \times (1 + j0)}{j(0.323 + 0.323 + 0.126)}$$

$$= -j3.886 \, \text{p.u. } \textbf{Ans.}$$

1.3   APPLICATION OF SOFT COMPUTING IN FAULT ANALYSIS

It is clear from the above examples that the conventional method of fault analysis requires the information of system parameters, fault location, and the type of fault. A smart protection system needs to act fast for which the identification of the type of fault and its location are necessary. Also, the design of protection system should be such that on occurrence of fault it should act independently with minimum input parameters.

Soft computation of Fault analysis has shown promising results as of now, and it involves mainly three major steps:

- Signal Processing
- Feature Extraction
- Analysis of Features by Training

Discrete data of any signal obtained from a system contain the information of the condition of that system. If the raw data is directly trained or used as input feature of a digital relaying circuit then the output may be erroneous due to the presence of noise. Hence, signal processing is necessary to obtain the characteristic features of any signal pertaining to a particular condition of a system. There are multiple signal processing tools available that have their own characteristics and applicable to specific kind of signals as follows:

1. Fast Fourier Transform (FFT)
2. Short Time Fourier Transform (STFT)
3. Wavelet Transform (WT)
4. S-Transform (ST)

The theory and mathematical details of these tools have been provided in Chapters 2 and 3.

1.4   COMPARISON OF CONVENTIONAL TECHNIQUES WITH SOFT COMPUTING

In both the methods of fault analysis, a single line diagram of the relevant power system network is necessary. The calculation of fault current by conventional method requires proper information of different system parameters, e.g., sequence impedance values of the transmission line, transformer, and sources. The fault location has to be specified without which the sequence diagrams/Thevenin's network cannot be drawn. The magnitude of fault impedance also affects the value of fault current but it is not an accessible parameter. The complications of the configurations of sequence networks increase with the increase in dimension of a power system network. High impedance fault currents are very difficult to calculate by conventional method because of the inaccessibility of the magnitude of fault impedance. The fault inception time and the duration of any fault are two important features that play an important role in determining the magnitude of fault current. These parameters are also inaccessible, and the conventional techniques do not have any scope of considering them.

Considering all these issues, soft computing methods provide a smart solution in which a fault condition can be simulated considering all the possible operating conditions. The resultant data of fault current, voltage, line voltage, and line current can be obtained, stored and subsequently analysed by a suitable signal processing method. Various statistical and non-statistical features can be calculated from the processed data. The features can be investigated and selected for training as input parameters by algorithms such as Artificial Neural Network, Support Vector Machine, and neuro-fuzzy technique. The output of such training algorithms provides the recognition of a particular fault condition and its location, irrespective of the magnitude of fault impedance, fault location, fault duration, and fault inception time.

The accuracy and speed of the entire method depend on the type of signal analysis tool, feature selection, number of samples of signal considered at any point of time, and the capacity of predictability of the training algorithm.

1.5 SUMMARY

This chapter provides emphasis on the emergence of soft computing techniques in power system fault analysis. The conventional methods have been discussed in brief. A comparison of the conventional methods with soft computing techniques has been provided with simple examples. It is clear that the conventional methods require the information of the system parameters, type of fault, and its location. Soft computation techniques have produced accurate and fast results with respect to fault identification, classification, and localisation under diverse operating conditions, and they require minimum input parameters for their functioning. The details of the applications of soft computing techniques have been elaborately explained in the subsequent chapters of this book.

REFERENCES

1. Nasser Tleis, *Power Systems Modelling and Fault Analysis, Theory and Practice*, 2nd Edition, Paperback ISBN: 9780128151174, eBook ISBN: 9780128151181, 2019, Elsevier.
2. Keith Harker, *High Voltage Power Network Construction*, Book doi: 10.1049/PBPO110E, Chapter doi: 10.1049/PBPO110E, ISBN: 9781785614231, e-ISBN: 9781785614248, IET.
3. Leonard L. Grigsby, *The Electric Power Engineering Handbook*, 3rd Edition, doi: 10.1201/b12111, 2012, CRC Press.
4. John Grainger, and William Stevenson, *Power System Analysis*, December 5, 2003, McGrawHill.

2 Wavelet Transform

2.1 INTRODUCTION

Mathematical transformation of a signal is required to obtain information that is not readily available from the raw signal. A raw signal generally implies its time–amplitude representation. A signal that is obtained after any mathematical transformation is known as a processed signal. There are multiple tools of mathematical transformation available out of which Fourier Transform (FT) is the most popular one. FT gives the information of the frequency content of any signal and this knowledge helps in diagnosis of a particular condition of any physical system. The frequency spectrum contains the magnitude of frequency components present in a signal. This information is sufficient for a signal in which the frequencies do not change with respect to time. Such a signal is known as a stationary signal. But in practice, the frequency components vary with respect to time and such a signal is known as non-stationary signal. In case of non-stationary signals, FT is not enough to recognize a particular condition of a system. It is also important to know the existence of frequency components with respect to time. Hence, Time-Frequency Representation (TFR) of a signal is needed. To achieve a TFR of any signal, a modified version of FT is available known as Short Time Fourier Transform (STFT).

WT converts a time-domain signal into the frequency domain in which the frequency components are represented with respect to time. Hence, WT provides a TFR of a signal. Unlike FT, WT uses a scalable window to accomplish the task of Time–Frequency conversion of a signal. WT is an excellent tool for the analysis of transient signals and image compression. Infinite functions are available in WT for the purpose of analysis. The main characteristic of these functions is that they have a finite time duration due to which the time–frequency localization of any signal is possible to be conducted. In the following sections, continuous wavelet transform (CWT) and discrete wavelet transform (DWT) have been described in brief.

2.2 BRIEF SURVEY OF LITERATURE

The method of Wavelet Transform (WT) was introduced in [1] where the transient signals in a power system network were analyzed through feature detection schemes. Another fast fault detection scheme was demonstrated in [2] where a combination of WT and a Neural Classifier was used for fault detection. An application of Morlet wavelets for the analysis of high-impedance fault generated signals is proposed in [3] and the superiority of WT over FFT was demonstrated in monitoring transient fault signals. The efficiency of WT in condition monitoring of electrical and mechanical machines was demonstrated in [4]. Another approach of high impedance fault detection scheme was proposed in [5] where the combination of DWT with frequency range and rms conversion produced a pattern recognition-based detection algorithm for electric distribution of high impedance fault detection. A fault classification scheme has been proposed in [6] where the fault detection and its clearing time are determined based on a set of rules obtained from the current waveform analysis in time and wavelet domains. The method is able to distinguish faults from other power-quality disturbances, such as voltage sags and oscillatory transients, which are common in power systems operation. An artificial neural network classifies the fault from the voltage and current waveforms pattern recognition in the time domain. The paper [7] describes the application of the adaptive whitening filter and the wavelet transform used to detect the abrupt changes in the signals recorded during disturbances in the electrical power network in South Africa. The main focus has been to estimate exactly the time instants of the changes in the signal model parameters during the pre-fault condition and following events like initiation of fault, circuit breaker opening, and auto-reclosure of the circuit breakers. A simple and effective algorithm has been proposed in [8] for detection of arcing faults in distribution networks where the phase displacement between wavelet coefficients is calculated for zero-sequence voltage and current signals at a chosen high-level frequency. The paper in [9] presents a procedure for determining fault location in MV distribution systems in which it has been established that a correlation exists between typical frequencies of the CWT-transformed signals and specific paths in the network covered by the traveling waves originated by the fault. The location of single line-to-ground faults in distribution lines has been obtained in [10] by the development of wavelet fuzzy neural network where the features have been extracted from the high-frequency components of the fault signals. A new technique based on DWT and ANN has been proposed in [11] for detection of high impedance faults in electrical distribution feeders. The paper in [12] deals with the application of wavelet transforms for the detection, classification, and location of faults on transmission lines. A Global Positioning System clock is used to synchronize sampling of voltage and current signals at both the ends of the transmission line. The detail coefficients of current signals of both the ends are utilized to calculate fault indices. These fault indices are compared with threshold values to detect and classify the faults. Artificial neural networks are employed to locate the fault, which make use of approximate decompositions of the voltages and currents of local end. A combination of DWT and back propagation neural network (BPNN) was used in [13] for fault detection and classification in parallel transmission lines. Different fault conditions were explored from which wavelet transform coefficients and wavelet energy coefficients were obtained as training features. A fault detection scheme has been suggested in [14] for a grid connected photovoltaic system based on WT. A methodology for fault classification and determination of its location has been suggested in [15] based on WT. A neuro-fuzzy classifier has been used to detect the fault type and the proposed method can also distinguish the power quality disturbances from the fault conditions in a power distribution network. A new application has been proposed in [16] which uses high-frequency content of a subset of local currents of one end of a protected line to classify transients on the line protected and its adjacent lines. The scheme can classify transients – including faults – occurring on a protected line, categorize transients on adjacent lines, and pinpoint the line causing the transient event. DWT is used to extract high-frequency components from the current signals. A feature vector is built using the wavelet details of coefficients to train an ANN. A novel simple and effective method of faulty feeder detection in resonant grounding distribution systems based on the continuous wavelet transform (CWT) and convolutional neural network (CNN) is presented in [17]. The time–frequency gray scale images are acquired by applying the CWT to the collected transient zero-sequence current signals of the faulty feeder and sound feeders. The features of the gray scale image will be extracted adaptively by the CNN, which is trained by a large number of gray scale images under various kinds of fault conditions and factors. The extraction of features and the faulty feeder detection can be implemented by the trained CNN simultaneously. A feature extraction method based on DWT has been proposed in [18] where the selected features have been trained by an evolving neural network to classify high impedance fault conditions. The proposed method has been compared with other established classifiers like SVM and PNN. The results have shown that the standard classifiers are efficient if the fault pattern do not change abruptly, whereas the evolving system is efficiently capable of fault detection even if there is any sudden change in fault condition. Wavelet packet transform (WPT) has been recently used in [19] to obtain energy coefficients from the current signals of the faulty phase of a power transmission network for detection of all kinds of possible fault conditions. Another version of WT, i.e., Maximal Overlap Discrete Wavelet Transform (MODWT) has been used in [20,21] for fault detection and classification in Indian Power System network where different attributes of the three phase current signals have been calculated like maximum change in wavelet energy and standard deviation values of MODWT coefficients. A new and fast approach of fault detection and classification has been proposed in [22] involving a combination of WT with Chebyshev Neural Network (ChNN). The suggested method uses only measured three-phase current signals at relaying end. The accuracy, speed, and effectiveness of the scheme have been verified with a fault data generation system developed on PSCAD/EMTP with different system parameter variations like fault resistance, load angle, fault inception angle, and types of faults. Recently, an evolving neuro-fuzzy network has been proposed in [23] for detection of high impedance faults where WPT has been used for feature extraction. The evolving neuro network is capable of recognizing other transient conditions as well and can also detect efficiently any abrupt change in the fault condition.

2.3 APPLICATION OF FT AND STFT IN STATIONARY AND NON-STATIONARY SIGNALS

FT is the most popular method of signal transformation in which a time-domain signal is converted into its frequency domain. The basic formulae of transformation of a time-domain signal $x(t)$ and the corresponding inverse transformation are given in Equations 2.3 and 2.2.

$$X(f) = \int_{-\infty}^{\infty} x(t) e^{-j2\pi ft} \, dt \tag{2.1}$$

$$x(t) = \int_{-\infty}^{\infty} X(f) e^{j2\pi ft} \, df \tag{2.2}$$

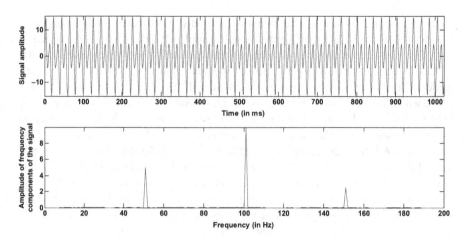

FIGURE 2.1 Waveform of Stationary signal given in Equation 2.3 and its frequency spectrum obtained from FFT.

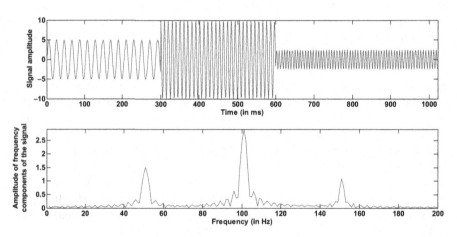

FIGURE 2.2 Waveform of non-stationary signal and its frequency spectrum obtained from FFT.

A stationary signal is defined by Equation 2.3 and the corresponding frequency spectrum is shown in Figure 2.1 after FFT. However, a non-stationary signal is also considered as shown in Figure 2.2 in which the same frequency components exist but at different instants of time. The frequency spectrum of the non-stationary signal is also shown along with the signal in Figure 2.2. Both the spectra look similar to each other except some glitches that appear in Figure 2.2 due to the changeover of magnitude of one frequency component to another at different instants of time. Hence, it is evident from this example that there is no way to obtain the information of frequency component with respect to time in FT.

$$y = 5\sin(2\pi ft) + 10\sin(2\pi 100t) + 2.5\sin(2\pi 150t) \quad (2.3)$$

The non-stationary signal shown in Figure 2.2 is defined as follows:

Interval, t1 = 0 to 300 ms, signal has a 50 Hz sinusoid;

t2 = 300 to 600 ms, signal has a 100 Hz sinusoid;

t3 = 600 to 1000 ms, signal has a 150 Hz sinusoid;

A sample program of FFT of stationary and non-stationary signals is shown below:

```
len=1024;
f=50;
N1=1;
N2=2;
N3=3;
t1=0:299;
    y1=5*sin(2*pi*N1*f*t1/len);
    t2= 300:599;
    y2=10*sin(2*pi*N2*f*t2/len);
    t3= 600:len-1;
    y3=2.5*sin(2*pi*N3*f*t3/len);

    y=[y1 y2 y3];
    t=[t1 t2 t3];

ts = 0:1:len-1;
ys=5*sin(2*pi*N1*f*ts/len)+10*sin(2*pi*N2*f*ts/
len)+2.5*sin(2*pi*3*f*ts/len);

    freqspectrum = abs(fft(y));
    harmonicamplitude = (2.*freqspectrum)/len;
    subplot(2,1,1), plot(t,y), axis([0 max(t) min(y)
max(y)]), xlabel('Time (in ms)'), ylabel (' Signal
amplitude')
   subplot(2,1,2), plot(harmonicamplitude) , axis([0 200 0
max(harmonicamplitude)]), xlabel('Frequency (in Hz)')
```

STFT on the other hand provides a TFR of a time-domain signal. Equation 2.4 below describes STFT of a time-domain signal $x(t)$ in which $\omega(t)$ is a window function and $\omega^*(t)$ is its complex conjugate. For every t' and f, a new STFT coefficient is computed.

$$STFT_x^{(\omega)}(t',f) = \int_t x(t)\omega^*(t-t')e^{-j2\pi ft}dt \qquad (2.4)$$

Hence both the stationary and non-stationary signals (considered earlier) have been subjected to STFT and the frequency spectra of both the signals are shown in Figure 2.3 and Figure 2.4, respectively.

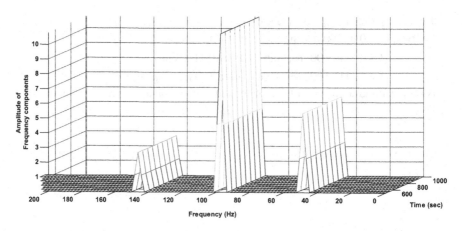

FIGURE 2.3   Waveform of stationary signal given in Equation 2.1 and its frequency spectrum obtained from STFT.

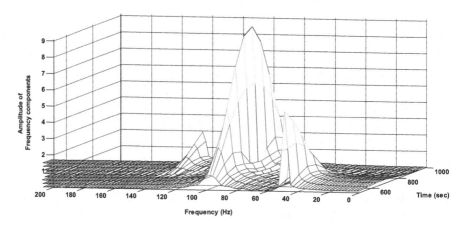

FIGURE 2.4   Waveform of non-stationary signal and its frequency spectrum obtained from STFT.

In Figure 2.4, the harmonic peaks are clearly visible at different instants of time but they overlap with each other with respect to time. Hence, the time resolution is poor and the frequency resolution is good. On the other hand, if we need good time resolution, then the frequency resolution will be poor. This happens due to the fixed width of the window function. Also, it has to be acknowledged that the time and resolution problem exists on the basis of the Heisenberg Uncertainty Principle irrespective of the type of transformation used.

A sample program in MATLAB is shown below:

```
clear, clc, close all
len=1024;
f=50;
```

```
N1=1;
N2=2;
N3=3;
t1=0:399;
     y1=5*sin(2*pi*N1*f*t1/len);
     t2= 400:699;
     y2=10*sin(2*pi*N2*f*t2/len);
     t3= 700:len-1;
     y3=2.5*sin(2*pi*N3*f*t3/len);
   x=[y1 y2 y3];
     t=[t1 t2 t3];
   ts = 0:1:len-1;
ys=5*sin(2*pi*N1*f*ts/len)+10*sin(2*pi*N2*f*ts/
len)+2.5*sin(2*pi*3*f*ts/len);
x = ys;
Fs = 1;
   %% STFT Parameters.
L  = length(x);
N  = 512; % Selected window size.
M  = 450; % Selected overlap between successive segments in
samples.
Nfft = 512; % Selected number of FFT points.
[t,f,S] = stft(x,N,M,Nfft,Fs,'hamm');
%% Plot the Spectrogram.
h = figure('Name','STFT - Method I Demo');
colormap('jet');
[T,F] = meshgrid(t,f*1000); % f in KHz.
%surface(T,F,10*log10(abs(S.^2) + eps),'EdgeColor','none');
mesh(T,F,(abs(S.*4/512)))
axis tight;
%grid on;
%title(['Signal Length: ',num2str(L),', Window Length: ',
num2str(N),', Overlap: ', num2str(M), ' samples.']);
xlabel('Time (sec)');
ylabel('Frequency (Hz)');
zlabel ('Amplitude of Frequency components');
```

The problem of resolution is addressed in multiresolution analysis (MRA) of signals in which the signal is analyzed at different frequencies with different resolutions. MRA is designed to give good time resolution and poor frequency resolution at high frequencies and good frequency resolution and poor time resolution at low frequencies. Most of the practical signals have high frequency components for short durations and low frequency components for long durations. MRA is achieved in wavelet transform which is described in the next function.

2.4   CONTINUOUS WAVELET TRANSFORM (CWT)

The following describes CWT in brief starting from its evolution. The detail description of wavelet is available in many literature studies and books [1–18]. An elaborate application of CWT has been provided after the description which would help the readers to apply CWT in processing of different types of time-domain signals to achieve the desired result.

2.4.1   EVOLUTION OF CWT FROM STFT

It is required to describe STFT in brief in order to understand CWT. STFT is a modified version of FT in which a time function is multiplied by a movable window function. The width of the window function is constant and it is shifted through the entire length of the signal. In FT, the time information is completely lost as the width of the window function exp(jwt) is infinite and it exists throughout the whole length of the signal. Unlike FT, there remains a problem of resolution in the case of STFT. Smaller the width of the window, better is the time resolution and poorer is the frequency resolution. The situation is vice versa in the case of wider window. This is illustrated with an example as shown below:

CWT is a modified version of STFT. In this method, a signal is multiplied by a window function similar to that in STFT. The difference is that the width of the window function is not constant and it changes by the variation of a scaling factor.

The basic equation of STFT of a time-domain function $x(t)$ is given by Equation 2.5.

$$STFT_x^{(\omega)}(t,f) = \int x(t)\omega(t-t')e^{-i2\pi ft}dt \qquad (2.5)$$

In Equation 2.5, the window function ω is fixed in width resulting in resolution problem. Henceforth, analysis with different resolutions at different frequencies was adopted in WT. In multiresolution transform like WT, a compromising situation is created in which good time resolution is obtained with poor frequency resolution at high frequencies. On the other hand, good frequency resolution is obtained with poor time resolution at low frequencies. Equation 2.4 defines the basic formula of WT, [1].

$$W(d,\tau) = \int_{-\infty}^{+\infty} f(t)\frac{1}{\sqrt{|d|}}\psi\left(\frac{t-\tau}{d}\right)dt \qquad (2.6)$$

In Equation 2.4, τ represents translation, i.e., the location of the window and d represents scale, i.e., the frequency parameter.

The window function is known as wavelet. A wavelet is a simple oscillatory function of finite duration. A mother wavelet is a wavelet that integrates to zero, has finite energy, and satisfies the admissibility condition. It is a prototype for generating other window functions. Examples are Morlet, Mexican Hat, etc. The equation of the mother wavelet is shown in Equation 2.7.

$$\Psi_{d,\tau}(t) = \frac{1}{\sqrt{|d|}} \psi\left(\frac{t-\tau}{d}\right) dt \qquad (2.7)$$

Different types of wavelet functions are available like, Haar wavelet, Daubechies family of wavelets (dbN), Meyer wavelet, and Morlet wavelet. In Daubechies family of wavelets db stands for Daubechies and N represents the order. The selection of the type of wavelet function depends on the nature of the signal to be processed.

2.5 APPLICATION OF CWT IN SIGNAL PROCESSING USING MATLAB

WT is an excellent technique of obtaining interesting features from any time-domain signal obtained from a physical system, which otherwise remain hidden and difficult to recognize.

This fact can be better explained by a case study as shown below:

A data set of three phase voltage signals has been considered which has been obtained from three-phase power system network during a line-ground fault simulated in Phase A. CWT has been initially implemented on all the signals. Each voltage signal has only 512 samples. Symlet 4 wavelet has been chosen in this example for analysis. It should be noted in this context that apart from Symlet a huge number of wavelets are available in the toolbox of CWT which the readers can apply depending on the nature of their signal. A number of wavelets can be tried to achieve the desired result. The size of the wavelet coefficient matrix generated after CWT is implemented in each signal is [64, 512], where 64 indicates the no. of coefficients and 512 represents the number of samples of the signal considered in the study.

The voltage signal of Phase A of a three-phase power system network during line-ground fault (Phase A shorted to ground) is shown in Figure 2.5. The fault signal has been simulated in MATLAB Simulink environment. The total time period of the waveform is 0.06 sec. The fault has been initiated at t = 0.04 sec, and it lasts for 0.2 sec.

CWT has been implemented on the volt age signal in MATLAB, and the corresponding commands are given below. The output of the program is a TFR which is shown in Figure 2.6

```
load c:\voltagedataB1\dataAG\dataAG100km\vdata1B1.txt;
COEFS1 = CWT(vdata1B1,1:64,'sym4','3Dplot');
```

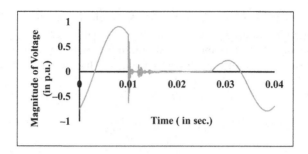

FIGURE 2.5   Voltage signal of Phase A of a three-phase power system network during line-ground fault (Phase A shorted to ground).

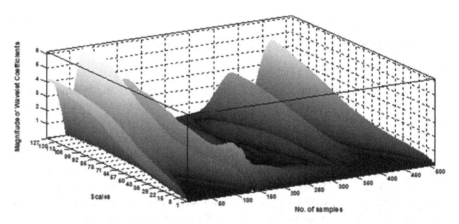

FIGURE 2.6   TFR of voltage signal of Phase A in case of AG fault obtained from CWT.

A matrix of wavelet coefficients is produced for each phase. The maximum value of the wavelet coefficient is calculated and termed as $\mathbf{C_{Fmax}}$. The same is calculated for the different types of fault locations in the transmission line. The magnitudes of the features for different fault locations have been plotted as shown in Figure 2.7 It is observed that during any fault condition, the magnitude of $\mathbf{C_{Fmax}}$ of the faulty phase is less than the healthy phases.

Please note that there are huge options available in selection of the type of features. The users will have to explore the matrix of wavelet coefficients to find them and judge their suitability in extraction of the desired information.

The magnitude of C_{Fmax} is used as input feature for training of neural classifier. Any other classifier can also be used. In this case, Probabilistic Neural Network (PNN) has been used. The theory of PNN has been explained in Chapter 6.

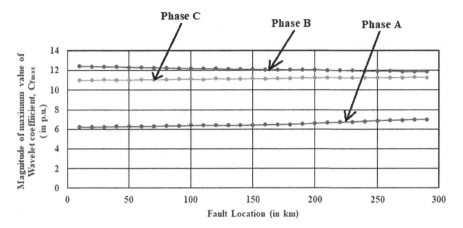

FIGURE 2.7   Plot of magnitude of **CFmax** for different fault locations in case of AG fault.

In MATLAB, the fault classification has been conducted by the application of PNN and the programming commands are shown below:

```
P = [P1 P2 P3 P4 P5 P6 P7 P8 P9 P10 P11]; ,P2,…P10 are the
set of input features for the 10 types of fault conditions
simulated in this case (It can be more than that), P11 is
the set of input features corresponding to the fault
condition
Tc = [x y z m n p q r s t u]; Tc is the Target vector
containing the integral values of the types of faults. For
e.g. x =1 for AG (Phase A shorted to ground) type of fault,
y=2 for BG (Phase B shorted to ground) type of fault and so
on.
T = ind2vec(Tc);
spread = 1;
net = newpnn(P,T,spread);
A = sim(net,P);
Ac = vec2ind(A);
p = [P1';P2';P3'];
a = sim(net,p);
ac = vec2ind(a)
```

The output of this program is an integral value from 1 to 11 given by the variable ac. **The explanation of each command is available in the help document of MATLAB software or any other book on MATLAB programming and hence, it is not given in this chapter**.

Similarly, some other features may be obtained from the matrix of wavelet coefficients to determine Fault location once the type of fault is identified. The readers are requested to find out the method of obtaining the fault location by using BPNN from the training of signal features that has been discussed in subsequent chapters.

2.6   DISCRETE WAVELET TRANSFORM (DWT)

The output coefficients of CWT are enormous and redundant. In DWT, scales and positions are based on powers of two, resulting in a lesser amount of coefficients. The scheme is implemented by using filters that were developed in 1988 by Mallat [24]. In this technique, a signal is decomposed into parts, approximations, and details. The approximations are the high-scale, low-frequency components of the signal. The details are the low-scale and high-frequency components. The decomposition process can be iterated, with successive approximations being decomposed in turn, so that one signal is broken down into many lower resolution components. This is known as Multilevel Decomposition and in this way a wavelet decomposition tree is generated, as shown in Figure 2.8. In reality, the decomposition can continue until the individual details consist of a single sample or pixel. In practice, the number of levels of decomposition would depend on the nature of the signal, the application in which DWT has been employed or on a suitable criterion such as entropy.

As an example, the same voltage waveform shown in Figure 2.5 during a line-ground fault occurring at Phase A of a single circuit power transmission network is considered here. DWT is implemented on the voltage signal of phase A and five levels of decomposition are shown in Figure 2.9. The magnitude of detail coefficients is shown in this figure as they are supposed to contain the relevant information of the signal.

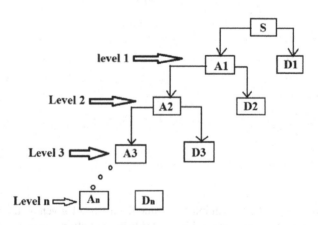

FIGURE 2.8   Wavelet decomposition tree.

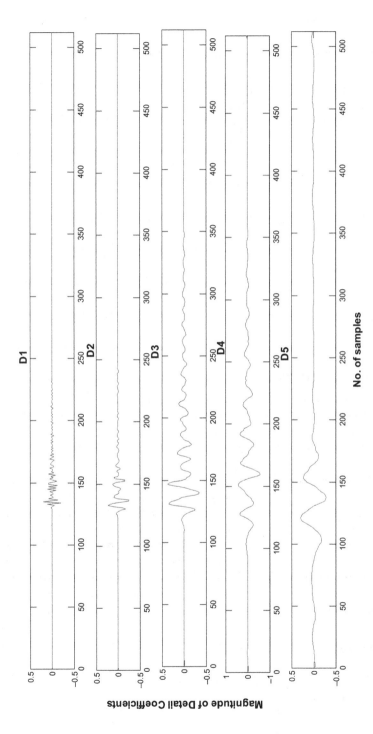

FIGURE 2.9 Plot of magnitudes of detail coefficients from levels 1–5 of the voltage waveform of Figure 2.5.

The programing in MATLAB is given as under:

```
load c:\voltagedataB1\dataAG\dataAG10km\vdata2B1.txt;
[C1,L1] = wavedec(vdata2B1,5,'db4');
A1 = wrcoef('a',C1,L1,'db4',1);
D1 = wrcoef('d',C1,L1,'db4',1);
A2 = wrcoef('a',C1,L1,'db4',2);
D2 = wrcoef('d',C1,L1,'db4',2);
A3 = wrcoef('a',C1,L1,'db4',3);
D3 = wrcoef('d',C1,L1,'db4',3);
A4 = wrcoef('a',C1,L1,'db4',4);
D4 = wrcoef('d',C1,L1,'db4',4);
A5 = wrcoef('a',C1,L1,'db4',5);
D5 = wrcoef('d',C1,L1,'db4',5);
subplot(5,1,1), plot(D1)
subplot(5,1,2), plot(D2)
subplot(5,1,3), plot(D3)
subplot(5,1,4), plot(D4)
subplot(5,1,5), plot(D5)
```

2.7 CONCLUSION

In this chapter, different signal processing techniques, namely, FFT, STFT, CWT, and DWT have been explained with the help of examples and case studies. It is being explained here that how CWT and DWT can be implemented on transient signals for feature extraction and the corresponding features can be analyzed by a suitable classifier for determination of a particular condition of any system. As the signal processing has been done in MATLAB, the sample programs have also been provided. The readers can apply the same programs on different types of signals with the necessary modifications according to their suitability. An elaborated comparative analysis can be conducted by the application of all the four methods on any transient/non-sinusoidal signal and important results can be obtained on the condition monitoring of any physical system. An extensive literature survey has been provided from which it is evident that CWT and DWT are efficient in the design of smart relaying system of a power system network, complex machinery, and condition monitoring of machines.

REFERENCES

1. D. C. Robertson, O. I. Camps, J. S. Mayer, and W. B. Gish, "Wavelets and electromagnetic power system transients," *IEEE Transactions on Power Delivery*, vol. 11, no. 2, pp. 1050–1056, 1996. doi: 10.1109/61.489367.

2. F. N. Chowdhury and J. L. Aravena, "A modular methodology for fast fault detection and classification in power systems," *IEEE Transactions on Control Systems Technology*, vol. 6, no. 5, pp. 623–634, 1998. doi: 10.1109/87.709497.

3. S. J. Huang and C. T. Hsieh, "High-impedance fault detection utilizing a Morlet wavelet transform approach," *IEEE Transactions on Power Delivery*, vol. 14, no. 4, pp. 1401–1407, 1999. doi: 10.1109/61.796234.

4. Z. K. Peng and F. L. Chu, "Application of the wavelet transform in machine condition monitoring and fault diagnostics: A review with bibliography," *Mechanical Systems and Signal Processing*, vol. 18, no. 2. pp. 199–221, 2004. doi: 10.1016/S0888-3270(03)00075-X.

5. T. M. Lai, L. A. Snider, E. Lo, and D. Sutanto, "High-impedance fault detection using discrete wavelet transform and frequency range and RMS conversion," *IEEE Transactions on Power Delivery*, vol. 20, no. 1, pp. 397–407, 2005. doi: 10.1109/TPWRD.2004.837836.

6. K. M. Silva, B. A. Souza, and N. S. D. Brito, "Fault detection and classification in transmission lines based on wavelet transform and ANN," *IEEE Transactions on Power Delivery*, vol. 21, no. 4, pp. 2058–2063, 2006. doi: 10.1109/TPWRD.2006.876659.

7. A. Ukil and R. Živanović, "Abrupt change detection in power system fault analysis using adaptive whitening filter and wavelet transform," *Electric Power Systems Research*, vol. 76, no. 9–10, pp. 815–823, 2006. doi: 10.1016/j.epsr.2005.10.009.

8. M. Michalik, W. Rebizant, M. R. Lukowicz, S. J. Lee, and S. H. Kang, "High-impedance fault detection in distribution networks with use of wavelet-based algorithm," *IEEE Transactions on Power Delivery*, vol. 21, no. 4, pp. 1793–1802, 2006. doi: 10.1109/TPWRD.2006.874581.

9. A. Borghetti, S. Corsi, C. A. Nucci, M. Paolone, L. Peretto, and R. Tinarelli, "On the use of continuous-wavelet transform for fault location in distribution power systems," *International Journal of Electrical Power and Energy Systems*, vol. 28, no. 9 SPEC. ISS., pp. 608–617, 2006. doi: 10.1016/j.ijepes.2006.03.001.

10. F. Chunju, K. K. Li, W. L. Chan, Y. Weiyong, and Z. Zhaoning, "Application of wavelet fuzzy neural network in locating single line to ground fault (SLG) in distribution lines," *International Journal of Electrical Power and Energy Systems*, vol. 29, no. 6, pp. 497–503, 2007. doi: 10.1016/j.ijepes.2006.11.009.

11. I. Baqui, I. Zamora, J. Mazón, and G. Buigues, "High impedance fault detection methodology using wavelet transform and artificial neural networks," *Electric Power Systems Research*, vol. 81, no. 7, pp. 1325–1333, 2011. doi: 10.1016/j.epsr.2011.01.022.

12. A. G. Shaik and R. R. V. Pulipaka, "A new wavelet based fault detection, classification and location in transmission lines," *International Journal of Electrical Power and Energy Systems*, vol. 64, pp. 35–40, 2015. doi: 10.1016/j.ijepes.2014.06.065.

13. A. Asuhaimi Mohd Zin, M. Saini, M. W. Mustafa, A. R. Sultan, "New algorithm for detection and fault classification on parallel transmission line using DWT and BPNN based on Clarke's transformation," *Neurocomputing*, vol. 168, pp. 983–993, 2015. doi: 10.1016/j.neucom.2015.05.026.

14. I. S. Kim, "On-line fault detection algorithm of a photovoltaic system using wavelet transform," *Solar Energy*, vol. 126, pp. 137–145, 2016. doi: 10.1016/j.solener.2016.01.005.

15. A. A. P. Bíscaro, R. A. F. Pereira, M. Kezunovic, and J. R. S. Mantovani, "Integrated fault location and power-quality analysis in electric power distribution systems," *IEEE Transactions on Power Delivery*, vol. 31, no. 2, pp. 428–436, 2016. doi: 10.1109/TPWRD.2015.2464098.

16. A. Abdullah, "Ultrafast transmission line fault detection using a DWT-based ANN," 2018. doi: 10.1109/TIA.2017.2774202.

17. M. F. Guo, X. D. Zeng, D. Y. Chen, and N. C. Yang, "Deep-learning-based earth fault detection using continuous wavelet transform and convolutional neural network in resonant grounding distribution systems," *IEEE Sensors Journal*, vol. 18, no. 3, pp. 1291–1300, 2018. doi: 10.1109/JSEN.2017.2776238.

18. S. Silva, P. Costa, M. Gouvea, A. Lacerda, F. Alves, and D. Leite, "High impedance fault detection in power distribution systems using wavelet transform and evolving neural network," *Electric Power Systems Research*, vol. 154, pp. 474–483, 2018. doi: 10.1016/j.epsr.2017.08.039.

19. A. R. Adly, R. El Sehiemy, M. A. Elsadd, and A. Y. Abdelaziz, "A novel wavelet packet transform based fault identification procedures in HV transmission line based on current signals," *International Journal of Applied Power Engineering (IJAPE)*, 2019. doi: 10.11591/ijape.v8.i1.pp11–21.

20. V. Ashok, A. Yadav, and A. Y. Abdelaziz, "MODWT-based fault detection and classification scheme for cross-country and evolving faults," *Electric Power Systems Research*, 2019. doi: 10.1016/j.epsr.2019.105897.

21. V. Ashok and A. Yadav, "A real-time fault detection and classification algorithm for transmission line faults based on MODWT during power swing," *International Transactions on Electrical Energy Systems*, 2020. doi: 10.1002/2050-7038.12164.

22. B. Y. Vyas, R. P. Maheshwari, and B. Das, "Versatile relaying algorithm for detection and classification of fault on transmission line," *Electric Power Systems Research*, 2020. doi: 10.1016/j.epsr.2020.106913.

23. S. Silva, P. Costa, M. Santana, and D. Leite, "Evolving neuro-fuzzy network for real-time high impedance fault detection and classification," *Neural Computing and Applications*, 2020. doi: 10.1007/s00521-018-3789-2.

24. *Wavelet Toolbox User's Guide*, Copyright 1997–2001 by The MathWorks, Inc.

3 Stockwell Transform

3.1 INTRODUCTION

Spectral analysis using Fourier transform (FT) has been a very popular technique of signal analysis. One drawback of FT is that it produces the amplitude and phase spectrum without the information of time. This is adequate for stationary time series in which the characteristics of the time series do not change with time. In non-stationary waveforms, the spectral content of the time series changes with time, and the time-averaged amplitudes/phases found by Fourier methods are inadequate to describe such phenomena. In 1946, Dennis Gabor [1] adapted the FT to analyze only a small section of the signal at a time – a technique called *windowing* the signal. Gabor's adaptation, called the *Short-Time Fourier Transform* (STFT), maps a signal into a two-dimensional function of time and frequency. On account of the fixed width of the window function used, STFT has a poor time frequency resolution. The wavelet transform (WT) on the other hand uses a window function which dilates and contracts with frequency. The WT does not retain the absolute phase information and the visual analysis of the timescale plots that are produced by the WT is intricate. A time frequency representation (TFR) developed by Stockwell [2], which combines the good features of STFT and WT, is called the Stockwell transform or S-transform. It can be viewed as a frequency-dependent STFT or a phase-corrected WT.

3.2 FAST FOURIER TRANSFORM (FFT), STFT, AND WT

ST is a hybrid of STFT and WT due to which a brief description of FFT, STFT, and WT is provided in the following subsections.

3.2.1 FFT AND STFT

The frequency spectrum of a signal shows what frequencies exist in the signal. The frequency content of a signal is measured by *Fourier transform* (FT). If the FT of a signal in time domain is taken, the frequency–amplitude representation of that signal is obtained. FT gives the frequency information of the signal, which means that it tells how much of each frequency exists in the signal. FT of a signal $x(t)$ is given by Equation (3.1)

$$X(f) = \int_{-\infty}^{\infty} x(t) e^{-i2\pi ft} dt \qquad (3.1)$$

Fast Fourier transform (FFT) is a method for computing discrete Fourier transform (DFT) with reduced execution time.

FT is unsuitable for signals with time varying frequencies, i.e., non-stationary signals because it gives no information of the time instant at which a particular frequency exists.

Equation (3.2) defines the STFT of a signal $x(t)$, [1].

$$STFT_x^{(\omega)}(t,f) = \int x(t)\omega(t-t')e^{-i2\pi ft}dt \tag{3.2}$$

It is seen in Equation (3.2) that the STFT of the signal is nothing but the FT of the signal multiplied by a window function, $\omega(t - t')$.

The problem with the STFT lies with the width of the window function that remains fixed for a particular window function [1]. On the basis of Heisenberg uncertainty principle, it is impossible to obtain exact time–frequency representation of a signal, i.e., one cannot know what spectral components exist at what instances of time. The TFR gives the time intervals in which certain band of frequencies exist, which is a resolution problem [1].

In FT there is no resolution problem in the frequency domain, i.e., the frequencies that exist in the signal can be exactly known. Similarly, there is no time resolution problem in the time domain, since the value of the original signal at every instant of time is known. Conversely, the time resolution in the FT and the frequency resolution in the time domain are zero. The perfect frequency resolution in the FT is due to the fact that the window used in the FT is its kernel, the exp{iwt} function, which lasts at all times from minus infinity to plus infinity. The window function used in STFT is of finite length and covers only a portion of the signal due to which the frequency resolution is weakened.

Hence, it can be concluded that

Wide window → good frequency resolution, poor time resolution.

Narrow window → good time resolution, poor frequency resolution.

3.2.2 WAVELET TRANSFORM (WT)

The continuous wavelet transform (CWT) was developed as an alternative approach to the STFT to overcome the resolution problem. The wavelet analysis is done in a similar way to the STFT analysis in the sense that the signal is multiplied with a function, similar to the window function in the STFT, and the transform is computed separately for different segments of the time-domain signal. However, there are two main differences between the STFT and the CWT [1,2]:

1. The FTs of the windowed signals are not taken, and therefore single peak will be seen corresponding to a sinusoid, i.e., negative frequencies are not computed.
2. The width of the window is changed as the transform is computed for every single spectral component, which is probably the most significant characteristic of the WT.

The CWT of a signal $f(t)$ is defined as follows:

$$W(s,\tau) = \int_{-\infty}^{+\infty} f(t) \frac{1}{\sqrt{|s|}} \psi\left(\frac{t-\tau}{s}\right) dt \tag{3.3}$$

As seen from Equation (3.3), the transformed signal is a function of two variables, τ and s, the translation and scale parameters, respectively. $\psi(t)$ is the transforming function, and it is called the mother wavelet. The term translation is used in the same sense as it was used in the STFT. It is related to the location of the window, as the window is shifted through the signal. This term, obviously, corresponds to time information in the transform domain. Unlike in STFT, WT uses a scale parameter, s.

The parameter scale in the wavelet analysis is similar to the scale used in maps. As in the case of maps, high scales correspond to a non-detailed global view (of the signal), and low scales correspond to a detailed view. Similarly, in terms of frequency, low frequencies (high scales) correspond to a global information of a signal (that usually spans the entire signal), whereas high frequencies (low scales) correspond to a detailed information of a hidden pattern in the signal (that usually lasts for a relatively short time). Scaling, as a mathematical operation, either dilates or compresses a signal. Larger scales correspond to dilated (or stretched out) signals and small scales correspond to compressed signals. Unlike the STFT which has a constant resolution at all times and frequencies, the CWT has a good time and poor frequency resolution at high frequencies, and good frequency and poor time resolution at low frequencies.

Calculating wavelet coefficients at every possible scale generates an awful lot of data. The enormity of data can be reduced by choosing a subset of scales and positions to make calculations. This is possible by choosing scales and positions based on powers of two, i.e., *dyadic* scales and positions. The resultant analysis will be much more efficient and just as accurate [2]. Such an analysis is known as discrete wavelet transform (DWT).

In DWT, a timescale representation of a digital signal is obtained using digital filtering techniques. In this case, filters of different cutoff frequencies are used to analyze the signal at different scales [2]. The signal is passed through a series of high pass filters to analyze the high frequencies, and it is passed through a series of low pass filters to analyze the low frequencies. Hence the signal (S) is decomposed into two types of components – *approximation* (A) and *detail* (D) [2]. The *approximation* is the high-scale, low-frequency component of the signal. The *detail* is the low-scale, high-frequency component. The decomposition process can be iterated, with successive approximations being decomposed in turn, so that one signal is broken down into many lower resolution components. This is called the *wavelet decomposition tree* which is shown in Figure 3.1 below. As decompositions are done on higher levels, lower frequency components are filtered out progressively. For example, level 3 approximate signal (A3) contains lower frequency components as compared to level 1 approximate signal (A1).

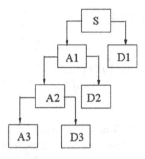

FIGURE 3.1   The Wavelet Decomposition Tree.

3.3   THEORY OF STOCKWELL TRANSFORM (ST)

The S-transform is an effectively efficient tool for TFR of a time series. ST is a hybrid of the STFT and WT, and it produces a frequency-dependent resolution with simultaneously localizing the real and imaginary spectra. It was first published in 1996 [3,4]. Due to its easy interpretation, multiresolution analysis, and the ability of maintaining the meaningful local phase information, the ST has established success in many areas including power quality, geophysics [5], and biomedicine [6,7].

3.3.1   DERIVATION OF ST FROM MODIFICATION OF STFT

In ST a particular window function known as normalized Gaussian function defined by Equation (3.4) is used by modifying Gaussian as a function of translation τ and dilation σ, as shown in Equation (3.5)

$$g(t,\sigma) = \frac{1}{\sigma\sqrt{2\pi}} e^{-\frac{t^2}{2\sigma^2}} \tag{3.4}$$

$$S^*(\tau,f,\sigma) = \int_{-\infty}^{\infty} h(t) \frac{1}{\sigma\sqrt{2\pi}} e^{-\frac{(t-\tau)^2}{2\sigma^2}} e^{-i2\pi ft} dt \tag{3.5}$$

For a particular value of σ Equation (3.5) is similar in definition to STFT shown in Equation (3.2). As Equation (3.5) is a function of three variables, it is simplified [3] by considering the width of the window σ to be inversely proportional to the frequency as shown in Equation (3.6).

$$\sigma(f) = \frac{1}{f} \tag{3.6}$$

Hence, ST is defined by [3,4,8–10],

$$S(\tau,f) = \frac{f}{\sqrt{2\pi}} \int_{-\infty}^{\infty} h(t) e^{-\frac{(t-\tau)^2 f^2}{2}} e^{-i2\pi ft} dt \tag{3.7}$$

The one-dimensional function of the time variable τ and fixed parameter $f1$ defined by $S(\tau, f1)$ is called a voice [4]. The one-dimensional function of the frequency variable f and fixed parameter $\tau1$ defined by $S(\tau1, f)$ is called a local spectrum. The S-transform can be written as a convolution [4] of two functions over the variable t,

$$S(\tau,f) = \int_{-\infty}^{\infty} p(t,f) g(t - \tau, f) dt \tag{3.8}$$

or,

$$S(\tau,f) = p(\tau,f) * g(\tau,f) \tag{3.9}$$

where

$$p(\tau,f) = h(\tau) e^{-2\pi i f \tau} \tag{3.10}$$

and

$$g(\tau,f) = \frac{f}{\sqrt{2\pi}} e^{-\frac{\tau^2 f^2}{2}} \tag{3.11}$$

Let $B(\alpha, f)$ be the FT (from τ to α) of the S-transform $S(\tau, f)$. By the convolution theorem, the convolution [4] in the τ (time) domain becomes a multiplication in the α (frequency) domain, as shown in Equation (3.12).

$$B(\alpha,f) = P(\alpha,f) G(\alpha,f) \tag{3.12}$$

where $P(\alpha, f)$ and $G(\alpha, f)$ are the FT of $p(\tau, f)$ and $g(\tau, f)$, respectively.

Hence, $B(\alpha, f)$ can be written as,

$$B(\alpha,f) = H(\alpha + f) e^{-\frac{2\pi^2 \alpha^2}{f^2}} \tag{3.13}$$

where $H(\alpha+f)$ is the FT of Equation (3.10) and the exponential term is the FT of the Gaussian function given by Equation (3.11). Hence, ST is the inverse FT (from α to τ) of Equation (3.13) (for $f \neq 0$).

$$S(\tau,f) = \int_{-\infty}^{\infty} H(\alpha + f) e^{-\frac{2\pi^2 \alpha^2}{f^2}} e^{i2\pi\alpha\tau} d\alpha \tag{3.14}$$

The exponential function in Equation (3.14) is the frequency-dependent localizing window and is called the Voice Gaussian [4].

The difference between ST and STFT is that the former gives better resolution in the case of high frequency components of a signal.

3.3.2   DERIVATION OF ST FROM MODIFICATION OF CWT

In order to utilize the information contained in phase of the CWT, it is necessary to modify the phase of the mother wavelet. The CWT, $W(\tau,d)$,of a function h(t) is defined as [3,9,10]:

$$W(\tau,d) = \int_{-\infty}^{\infty} h(t)w(t-\tau,\ d)dt \qquad (3.15)$$

where $W(\tau,d)$ is a scaled replica of the fundamental mother wavelet; the dilation determines the width of the wavelet, and this controls the resolution. The ST is obtained by multiplying the CWT with a phase factor, as defined below [3,4]:

$$S(\tau,f) = e^{i2\pi f\tau}W(d,\tau) \qquad (3.16)$$

where the mother wavelet for this particular case is defined as:

$$w(t,f) = \frac{|f|}{\sqrt{2\pi}}e^{-\left(\frac{t^2 f^2}{2}\right)}e^{-i2\pi ft} \qquad (3.17)$$

In Equation (3.15) dilation factor d is the inverse of frequency f. Thus, the final form of the continuous S-transform is obtained as

$$S(\tau,f) = \int_{-\infty}^{\infty} h(t).\left(\frac{|f|}{\sqrt{2\pi}}\right)e^{\left(-\frac{(\tau-t)^2 f^2}{2}\right)}e^{(-i2\pi ft)}dt \qquad (3.18)$$

and width of the Gaussian window is

$$\infty(f) = T = 1/|f| \qquad (3.19)$$

Since S-transform is a representation of local spectra, Fourier or time average spectrum can be directly obtained by averaging local spectra through inverse S-transform, as given by Equation (3.20) below:

$$h(t) = \int_{-\infty}^{\infty}\left\{\int_{-\infty}^{\infty} S(\tau,f)d\tau\right\}e^{i2\pi ft}df \qquad (3.20)$$

Equation (3.18) is a phase correction of the definition of the WT, [4,10,11] in which the mother wavelet is separated into two parts, the slowly varying envelope (the Gaussian function) which localizes in time, and the oscillatory exponential kernel $e^{-i2\pi ft}$ which selects the frequency being localized. It is the time localizing Gaussian that is translated while the oscillatory exponential kernel remains stationary. By not translating the oscillatory exponential kernel, the S-transform localizes the real and the imaginary components of the spectrum independently, localizing the phase spectrum as well as the amplitude spectrum. This is referred to as absolutely referenced phase information [4].

3.4   DISCRETE S-TRANSFORM (DST)

An electrical signal h(t) can be expressed in discrete form as $h(kT)$, k = 0, 1, ..., N-1 and T is the sampling time interval, [3,4].

The discrete FT of $h(kT)$ is obtained as,

$$H\left[\frac{n}{NT}\right] = \frac{1}{N}\sum_{k=1}^{N-1} h(kT)e^{\frac{-i2\pi k}{N}} \qquad (3.21)$$

where n = 0, 1, ..., N-1.

Using (4), the ST of a discrete time series is obtained by letting $f \rightarrow n/NT$ and $\tau \rightarrow jT$ as

$$S\left[jT,\frac{n}{NT}\right] = \sum_{m=0}^{N-1} H\left[\frac{m+n}{NT}\right]G(m,n)e^{i2\pi mj/N} \qquad (3.22)$$

and $G(m,n) = e^{-2\pi2m2/n2}$, $n \neq 0$ where j, m = 0,1, 2, ...N-1 and n = 1, 2, ...N-1

Equation (3.22) generates a complex matrix, the rows of which are the frequencies, whereas the columns are the corresponding times. The amplitude of the ST spectrum is obtained from the absolute values of the complex matrix. Each column, thus, represents the local spectrum at one point in time. The matrix preserves the amplitude information of the frequency content of the signal at different resolutions.

3.5   PROPERTIES OF ST

This section categorically presents the main properties of ST.
- **Inverse of the S-Transform and the FT**: As ST is a representation of the local spectrum, the average of the local spectra over time gives the FT spectrum.

$$\int_{-\infty}^{\infty} S(\tau,f)d\tau = H(f) \qquad (3.23)$$

where $H(f)$ is the FT of $h(t)$.

$h(t)$ can be recovered from $S(\tau,f)$ by Equation (3.24),

$$h(t) = \int_{-\infty}^{\infty}\left\{\int_{-\infty}^{\infty} S(\tau,f)d\tau\right\}e^{i2\pi ft}df \qquad (3.24)$$

Hence, ST is a generalization of FT to non-stationary time series.

- **Progressive resolution of the time–frequency domain**: The sampled frequencies of a time series consisting of N points with a sampling interval of T are $fn = n/(NT)$. Thus the periods sampled range from NT for the first harmonic (ignoring the DC level), $NT/2$ for the second harmonic, and so on as a function of $1/n$.

    Hence, the difference in time period between the second and third harmonic is $= NT/3 - NT/2 = NT/6$ and that between 62nd and 63rd harmonic is $= NT/62 - NT/63 = 2.56e-4NT$.

    Hence, the difference in time period between the high-frequency components becomes very smaller due to which it is difficult to distinguish high-frequency signals from each other.

    In other words, low-frequency signals have good frequency resolution and poor time resolution. The frequency resolution is poor for high-frequency signals but they have a good time resolution. This relationship between frequency and time resolutions follows the Heisenberg uncertainty principle.

- **Frequency Resolution depends on the signal**: In the case of ST, the frequency resolution is proportionally related to the length of the signal. The frequency resolution improves if the signal lasts for a longer period of time. The increase in the number of samples of the time-domain signal within the same period gives a better resolution of the harmonic components in the amplitude spectrum.

- **The S-Transform and Generalized Instantaneous Frequency**: The output of ST is a complex matrix. A particular voice of the S-transform is defined as [4],

$$S(\tau,f_o) = A(\tau,f_o)e^{i\phi(\tau,f_o)} \tag{3.25}$$

where

$$A(\tau,f_o) = \sqrt{R\{S(\tau,f_o)\}^2 + I\{S(\tau,f_o)\}^2} \tag{3.26}$$

and

$$\phi(\tau,f_o) = \arctan\left(\frac{I\{S(\tau,f_o)\}}{R\{S(\tau,f_o)\}}\right) \tag{3.27}$$

Since a voice isolates one specific component, the phase in Equation (3.27) is used to determine the instantaneous frequency in Equation (3.28).

$$IF(\tau,f_o) = \frac{1}{2\pi}\frac{\partial}{\partial\tau}\{2\pi\tau f_o + \phi(\tau,f_o)\} \tag{3.28}$$

- **Linearity**: The S-transform is a linear operation on the time series h(t). Equation (3.29) shows the model of a noisy signal.

$$data(t) = signal(t) + noise(t) \tag{3.29}$$

The linear property of ST leads to

$$S(data) = S(signal) + S(noise) \tag{3.30}$$

This is an advantage over the bilinear class of TFRs, where such linear relation does not exist as shown in Equation (3.30).

3.6   COMPARISON OF ST WITH CWT

As ST is an extension of CWT, major differences between ST and CWT have been summarized in this section.

1. **Frequency and time sampling**: Sampling of the frequency space in DST is identical to that of DFT. Simultaneously, DST retains the sampling of the time series. On the other hand, WT has a loosely defined scaling. It normally employs an octave scaling for frequencies, which results in an oversampled representation at the low frequencies and an undersampled representation at the higher frequencies.
2. **Direct Signal Extraction**: The amplitude, frequency, and phase at any time instant can be directly measured from the S-transform. This section is described in detail in Chapter 3 with examples. This direct extraction of a signal is due to the combination of absolutely referenced phase information and frequency invariant amplitude of the ST, and such direct extraction cannot be done with wavelet methods.
3. **ST Phase**: The ST retains the absolute phase information, whereas the phase information is lost in the WT. The absolutely referenced phase of the S-transform is in contrast to a wavelet approach, where the phase of the WT is relative to the center (in time) of the analyzing wavelet. Thus, as the wavelet translates, the reference point of the phase translates. In ST, the sinusoidal component of the basis function remains stationary, while the Gaussian envelope translates in time. Thus, the reference point for the phase remains stationary.
4. **ST Amplitude**: The unit area localizing function (the Gaussian) preserves the amplitude response of the S-transform and ensures that the amplitude response of the ST is invariant to the frequency. Just as the phase of the ST means the same as the phase of the FT, the amplitude of the ST means the same as the amplitude of the FT. On the other hand, WT diminishes the higher-frequency components.

3.7   SUMMARY

A concept of standard ST has been presented in this chapter. ST is a modified version of STFT or an extension of WT. It is based on a scalable localizing Gaussian window and supplies the frequency-dependent resolution. The WT is good at extracting information from both time and frequency domains. However, the WT is sensitive to noise. The properties of ST are that it has a frequency-dependent resolution of time–frequency domain and entirely refer to local phase information. ST is widely used in various applications such as heart sound analysis, image watermarking, filter design, seismogram analysis, analysis of engine induction noise in acceleration [8], and power quality analysis [10–17].

One main drawback of the ST is the amount of information redundancy in its resulting time–frequency representation. That causes large computing consumption and limits its use in dealing with large size of data. However, as a practical consideration, a long time series can be decomposed into multiple sections according to the convenience of the user [4], the resultant of which would be a S-matrix for each section. This would save the computational time and the further analysis of the smaller size S-matrices would be relatively more convenient. Additionally, to improve the computational efficiency of ST, the discrete orthonormal Stockwell transform (DOST) was proposed [18,19]. A computationally fast version of discrete ST (DST) was presented in [20] where cross-differential protection scheme for power transmission systems has been proposed. However, the scope of this book has been limited to the application of standard ST.

REFERENCES

1. https://ccrma.stanford.edu/~unjung/mylec/WTpart1.html and https://ccrma.stanford.edu/~unjung/mylec/WTpart2.html
2. M. Misiti, Y. Misiti, G. Oppenheim, and Jean-Michel Poggi, *Wavelet Toolbox User's Guide*, Version 2.1, COPYRIGHT 1997–2001 by The MathWorks, Inc.
3. F. Zhao and R. Yang, "Localization of the complex spectrum: S-transform", *IEEE Transactions on Signal Processing*, vol. 44, no. 4, pp. 998–1001, April 1996.
4. R. G. Stockwell, *S-Transform Analysis of Gravity Wave Activity from a Small Scale Network of Airglow Imagers*, PhD thesis, University of Western Ontario, London, Ontario, Canada, 1999.
5. M. G. Eramian, R. A. Schincariol, L. Mansinha, and R. G. Stockwell, "Generation of aquifer heterogeneity maps using two-dimensional spectral texture segmentation techniques", *Mathematical Geology*, vol. 31, pp. 327–348, 1999.
6. B. G. Goodyear, H. Zhu, R. A. Brown, and J. R. Mitchell, "Removal of phase artifacts from fMRI data using a Stockwell transform filter improves brain activity detection", *Magnetic Resonance in Medicine*, vol. 51, pp. 16–21, 2004.
7. C. Liu, W. Gaetz, and H. Zhu, "Estimation of time-varying coherence and its application in understanding brain functional connectivities", *EURASIP Journal on Advances in Signal Processing*, 2010, 390910. https://doi.org/10.1155/2010/390910
8. Yu-Hsiang Wang, *The Tutorial: S Transform*, Graduate Institute of Communication Engineering, National Taiwan University, Taipei, Taiwan, ROC.
9. C. R. Pinnegar and L. Mansinha, "The S-transform with windows of arbitrary and varying shape", *Geophysics*, vol. 68, no. 1, pp. 381–385, Jan–Feb 2003.
10. J. B. Reddy, D. K. Mohanta, and B. M. Karan, "Power system disturbance recognition using wavelet and S-transform techniques", *International Journal of Emerging Electric Power Systems*, vol. 1, no. 2, Article 1007, 2004.
11. F. Zhao and R. Yang, "Power-quality disturbance recognition using S-transform", *IEEE Transitions on Power Delivery*, vol. 22, no. 2, pp. 944–950, Apr 2007.

12. P. K. Dash, B. K. Panigrahi, and G. Panda, "Power Quality Analysis using S-Transform", *IEEE Transitions on Power Delivery*, vol. 18, no. 2, pp. 406–411, Apr 2003.

13. M. V. Chilukuri and P. K. Dash, "Multiresolution S-transform-based fuzzy recognition system for power quality events", *IEEE Transitions on Power Delivery*, vol. 19, no. 1, pp. 323–330, Jan 2004.

14. F. Zhao and R. Yang, "Power-quality disturbance recognition using S-transform", *IEEE Transitions on Power Delivery*, vol. 22, no. 2, pp. 944–950, Apr 2007.

15. S. Mishra, C. N. Bhende, and B. K. Panigrahi, "Detection and classification of power quality disturbances using S-transform and probabilistic neural network", *IEEE Transitions on Power Delivery*, vol. 23, no. 1, pp. 280–287, Jan 2008.

16. Ameen M. Gargoom, Nesimi Ertugrul, and Wen. L. Soong, "Automatic classification and characterisation of power quality events", *IEEE Transitions on Power Delivery*, vol. 23, no. 4, pp. 2417–2425, Oct 2008.

17. S. Hasheminejad, S. Esmaeili, and S. Jazebi, "Power quality disturbance classification using S-transform and hidden Markov model", *Electric Power Components and Systems*, vol. 40, no. 10, pp. 1160–1182, 2012.

18. R. G. Stockwell, "A basis for efficient representation of the S-Transform", *Digital Signal Processsing*, vol. 17, pp. 371–393, 2007.

19. Y. Wang and J. Orchard, "Fast discrete orthonormal Stockwell transform, SIAM", *Journal of Science Computation*, vol. 31, no. 5, pp. 4000–4012, 2009.

20. K. R. Krishnanand, and P. K. Dash, "A new real-time fast discrete S-transform for cross-differential protection of shunt-compensated power systems", *IEEE Transactions Power Delivery*, vol. 28, no. 1, pp. 402–410, 2013.

4 Application of ST for Time Frequency Representations (TFRs) of Different Electrical Signals

4.1 INTRODUCTION

The use of solid-state devices and equipment is increasing rapidly in electrical systems, and they introduce harmonics in the system. As a result, the signals in electrical systems deviate from being a pure sinusoidal one. Analysis of these non-sinusoidal signals is mainly accomplished using various kinds of signal analysis tools [1–3], like Short-Time Fourier Transform (STFT) and Wavelet Transform (WT). ST being a modified version of STFT and WT can be conveniently applied for TFR analysis of non-sinusoidal signals. ST provides the necessary information (amplitude, phase angle) if the data of the time series of any signal is available, and an equation of the corresponding signal can be framed.

This chapter illustrates the TFR analysis of different non-sinusoidal waveforms. Section 4.2 presents the application of ST in framing of equation of an electrical signal. A comparative analysis of FFT, DWT, and ST in TFR of stationary and non-stationary signals has been presented in Section 4.3. The effect of noise in TFR of electrical signals has been demonstrated in Section 4.4. A summary of the comparison of all the three signal processing methods has been tabulated in Section 4.5

4.2 SIGNAL EXTRACTION FROM A TIME SERIES

This section illustrates the method of framing equation of a signal from the data of time series. Non-sinusoidal signals have been simulated in MATLAB and the resultant data of the discrete time series have been analyzed by both FFT [4,5] and discrete ST (DST) [6–8].

4.2.1 NON-SINUSOIDAL WAVEFORM WHOSE EQUATION IS KNOWN

The Fourier series equation of the waveform simulated in MATLAB is given below, and the plot of the voltage waveform is shown in Figure 4.1 for only 200 samples. The fundamental frequency of the voltage waveform is 50 Hz, and a total of 1024 sample points of the voltage waveform are taken for analysis.

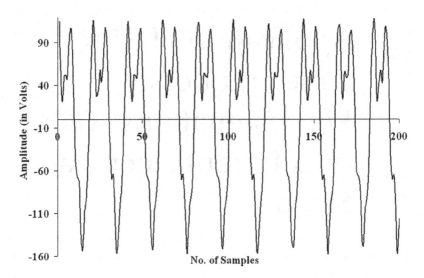

FIGURE 4.1   The complex voltage waveform.

$$v(t) = 100\sin\left(\omega t + 30^0\right) + 50\cos\left(2\omega t + 60^0\right)$$
$$+ \ 25\sin\left(3\omega t + 120^0\right) + 20\sin\left(5\omega t + 150^0\right) \qquad (4.1)$$
$$+ \ 10\sin\left(7\omega t + 50^0\right) + \dots\dots\dots\dots\dots\dots\dots$$

Both ST and FFT have been employed on the time series of the voltage equation. Figure 4.2 shows the frequency spectra obtained from ST and FFT. It is observed that there are sharp peaks at fundamental frequencies of 50 Hz and 100 Hz. There are small peaks at 150 Hz, 250 Hz, and 350 Hz. The amplitudes of the peaks are given in Table 4.1. In Figure 4.2 (b) sharp peaks at the frequencies of 50 Hz, 100 Hz, 150 Hz, 250 Hz, and 350 Hz are clearly visible in the plot. Hence, it is evident from both Figures 4.2 (a) and (b) that the main frequency components of the voltage waveform apart from the fundamental are the second, third, fifth, and seventh harmonics. This conclusion agrees well with the Equation (4.1). The amplitudes of the harmonics can be obtained from the absolute values of the S-matrix (STA matrix). Table 4.1 gives the comparison between the actual values of the magnitudes of the harmonics (obtained from the Equation 4.1) and those obtained from the STA matrix.

From Table 4.1 it is observed that the **absolute value of** maximum percentage error is 0.318%. Table 4.2 shows the magnitudes of the phase angles of the harmonic components obtained from the argument of the S-matrix.

The information of the amplitude and phase angle of each harmonic component obtained from ST can be used for framing the equation of the waveform as shown below:

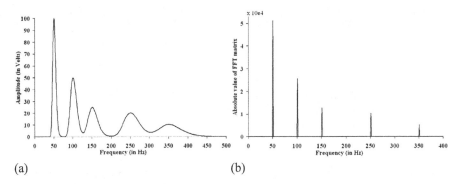

(a)   (b)

FIGURE 4.2   (a) Frequency spectrum of waveform in Figure 4.1 obtained from S-transform and (b) frequency spectrum of waveform in Figure 4.1 obtained from FFT.

TABLE 4.1
Comparison between the Magnitudes of the Harmonic Components Obtained from Equation (4.1) and the STA Matrix

Order of Harmonic	Peak Values of Harmonics		% Error (Absolute Value)
	From Equation (3.1)	From STA Matrix	
Fundamental component	100	100	0
Second Harmonic	50	50	0
Third	25	25.0077	0.0308
Fifth	20	20.0001	0.0005
Seventh	10	10.0318	0.318

TABLE 4.2
Magnitudes of the Absolute Values of the Phase Angles Obtained from the STA Matrix

Order of Harmonic	Phase Angle (in degree)
Fundamental component	60
Second Harmonic	60
Third	30
Fifth	60
Seventh	40

$$
\begin{aligned}
vnew(t) &= 100\cos\left(\omega t - 60^{0}\right) + 50\cos\left(2\omega t + 60^{0}\right) + \\
&\quad + 25.0077\cos\left(3\omega t + 30^{0}\right) + 20.0001\cos\left(5\omega t + 60^{0}\right) \\
&\quad + 10.0318\cos\left(7\omega t - 40^{0}\right) + \dots\dots\dots\dots \\
&\approx 100\sin\left(\omega t + 30^{0}\right) + 50\cos\left(2\omega t + 60^{0}\right) \\
&\quad + 25.01\sin\left(3\omega t + 120^{0}\right) + 20\sin\left(5\omega t + 150^{0}\right) \\
&\quad + 10.03\sin\left(7\omega t + 50^{0}\right) + \dots\dots\dots\dots
\end{aligned}
\tag{4.2}
$$

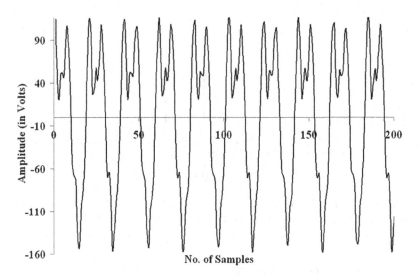

FIGURE 4.3   The plot of the waveform defined by Equation (4.2).

Figure 4.3 shows the plot of the waveform obtained from Equation (4.2).

The waveforms shown in Figures 4.1 and 4.3 are almost identical to each other. Hence, ST can be effectively employed for obtaining the magnitudes and phase angles of the harmonic components of a non-sinusoidal signal, using which the equation of the waveform can be written. In the following section the same method has been applied to determine the equation of magnetic inrush current waveform of a transformer.

4.2.2   HARMONIC ANALYSIS OF THE INRUSH CURRENT WAVEFORM OF A SATURATED TRANSFORMER

The inrush current waveform of a saturated transformer has been simulated in MATLAB Simulink environment. Figure 4.4 shows the circuit in which one phase of a three-phase transformer is connected to a 500 kV, 5000 MVA network. The transformer is rated at 500 kV/230 kV, 450 MVA (150 MVA per phase). The fundamental frequency of the system is 50 Hz. The flux–current saturation characteristic of the transformer is modeled with the hysteresis. A three-phase programmable voltage source is used to vary the internal voltage of the equivalent 500 kV network. During the first three cycles, source voltage is programmed at 0.8 p.u. Then, at t = 3 cycles (0.06 s), the voltage is increased by 37.5% (up to 1.10 p.u.).

In order to illustrate remnant flux and inrush current at transformer energization, the circuit breaker which is initially closed is first opened after six cycles, and then it is reclosed after nine cycles. The initial flux in the transformer is set at zero and source phase angle is adjusted at 90° so that flux remains symmetrical around zero when simulation is started. The inrush current is a transient signal and its amplitude will decrease with time, and at steady state, it will reduce to the value of no-load current of the transformer. Hence, a part of the inrush current waveform is chosen for

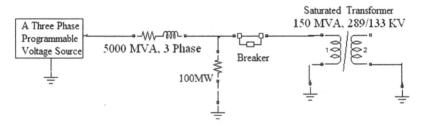

FIGURE 4.4   The simulated system in MATLAB considered for analysis.

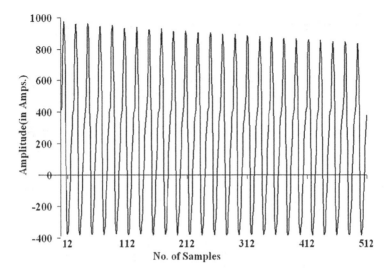

FIGURE 4.5   Inrush-current waveform of a saturated transform.

analysis within which the amplitude of the signal will almost remain constant. Figure 4.5 shows the plot of inrush-current waveform for 512 samples.

Figure 4.6 shows the frequency spectra obtained from both ST and FFT. It is observed from the Figures 4.6(a) and (b) that the inrush current has a d.c. component along with a pronounced second harmonic and a small amount of third harmonic. The amplitudes of the harmonics and the absolute values of the phase angles of the harmonic components are obtained from the S-matrix and are tabulated in Table 4.3.

Using the data from Table 4.3, the equation of the current waveform has been framed below:

$$i(t) \simeq 224 + 564.78 \cos\left(\omega t - 45^0\right) +$$
$$224.42 \cos\left(2\omega t + 179.9^0\right) + \tag{4.3}$$
$$72.09 \cos\left(3\omega t + 90.7^0\right) + \ldots\ldots$$

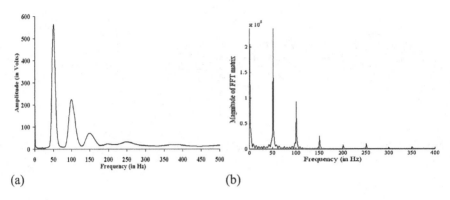

(a)   (b)

FIGURE 4.6   (a) Frequency spectrum of inrush current waveform obtained from S-transform and (b) frequency spectrum of inrush current waveform obtained from FFT.

TABLE 4.3

Magnitudes of the Harmonic Components and the Absolute Values of the Phase Angles Obtained from the STA Matrix

Order of Harmonic	Amplitude (in Amp.)	Phase Angle (in degree)
DC component	225.0026	0
Fundamental component	564.7794	45
Second Harmonic	224.423	179.9
Third Harmonic	72.0865	90.7

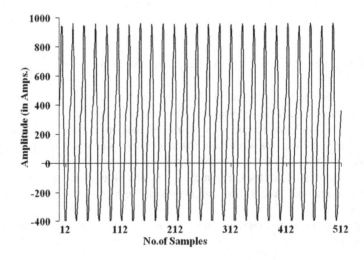

FIGURE 4.7   Inrush current waveform of a saturated transform.

The plot of the current equation $i(t)$ is shown in Figure 4.7.

The waveforms of Figures 4.5 and 4.7 appear similar to each other. The maximum deviation in the peak values in the positive half is −8.7% and that in the negative half is −8.2%. The percentage error is high compared to the previous cases. This is mainly due to the fact that the transient signal under study has been resolved into a number of small parts and each such part has been considered as a steady state signal, where the change in magnitude of the signal has not been considered. However, it can be concluded that an accurate equation can be framed of any non-sinusoidal signal and an approximate equation can be determined by any transient signal.

4.3 COMPARISON AMONG FFT, DWT, AND DST

Two types of signals: stationary and non-stationary are simulated in MATLAB. DWT [9–13], FFT, and ST have been employed on the signals and the corresponding outputs have been compared.

4.3.1 STATIONARY SIGNAL

A stationary signal has been simulated in MATLAB as shown in Figure 4.8(a), and its modeling equation is given in Equation (4.4). The fundamental frequency of the voltage waveform is 50 Hz, and a total of 1024 sample points of the voltage waveform have been considered for the present analysis.

$$\begin{aligned} v(t) &= 100\cos\left(\omega t - 60^0\right) + 50\cos\left(2\omega t + 60^0\right) \\ &+ 25\cos\left(3\omega t + 30^0\right) + 20\cos\left(5\omega t + 60^0\right) \\ &+ 10\cos\left(7\omega t - 40^0\right) + \ldots\ldots \end{aligned} \tag{4.4}$$

DWT, FFT, and ST have been employed on the same signal, and the corresponding outputs have been plotted in Figure 4.8.

Figure 4.8(b) shows the plot of the amplitude spectrum obtained from FFT. It is observed in Figure 4.8(b) that there are sharp peaks at the frequencies of 50 Hz, 100 Hz, 150 Hz, 250 Hz, and 350 Hz.

S-transform has been employed on the time series of the voltage equation. From the STA matrix, the frequencies and the amplitudes of the harmonic components have been obtained. The amplitude spectrum from ST has been shown in Figure 4.8(c) and sharp peaks are visible at the frequencies of 50 Hz and 100 Hz. Smaller peaks appear at 150 Hz, 250 Hz, and 350 Hz. The amplitudes of the peaks are given in Table 4.4. Hence, it is evident from both the Figures 4.8 (b) and 4.8 (c) that the main frequency components of the voltage waveform apart from the fundamental are the second, third, fifth, and seventh harmonics. This conclusion agrees well with the Equation 4.4.

The phase angle of the respective frequency components has been obtained from the argument of both FFT and ST matrix. The values are shown in Table 4.4. It is observed from Figure 4.8(d) that the phase angle obtained from the argument of S-matrix for every frequency component remains stationary with time.

DWT has been employed on this signal using the Daubechies4 mother wavelet. Figure 4.8 (e) shows the DWT coefficients at level 2. The periodic nature of the multiple frequency components is observed in Figure 4.8(e).

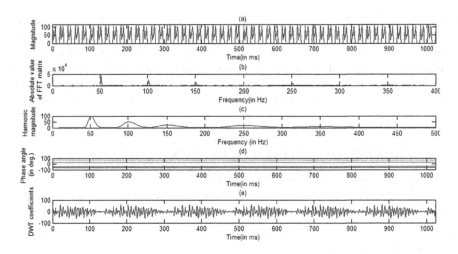

FIGURE 4.8 (a) Original signal, (b) Fourier spectrum of pure sinusoidal signal obtained from FFT, (c) amplitude spectrum of the same signal obtained from ST, (d) phase angle of the frequency components vs no. of samples, obtained from ST, and (e) magnitude of DWT coefficients at level 2 of the same signal.

TABLE 4.4

Magnitudes of Harmonic Components and Phase Angle Obtained from ST and FFT

TFR Method	ST		FFT	
Frequency (in Hz)	Amplitude (in Volts)	Phase Angle (in degrees)	Amplitude of FFT Matrix (× 10e4)	Phase Angle (in degrees)
50	100	−60	5.12	−60
100	50	60	2.56	60
150	25	30	1.28	30
250	20	60	1.02	60
350	10	−40	0.51	−40

Both ST and FFT generate complex values from which the phase angle of the frequency components can be obtained accurately. Hence, for stationary non-sinusoidal signal FFT is relatively more suitable than ST. It is seen from Table 4.4 that the amplitude of the frequency components can be directly obtained from STA matrix.

4.3.2 Non-Stationary Signal

DWT and ST are applicable for analyzing non-stationary signals. A non-stationary signal is simulated in MATLAB and is shown in Figure 4.9(a). It has four different frequency components occurring at different intervals of time.

In Figure 4.9 (b), it is clearly observed that the signal has four different frequency components. But there is no information about whether the frequencies are occurring at all intervals of time or at different time intervals. FFT fails to give this information.

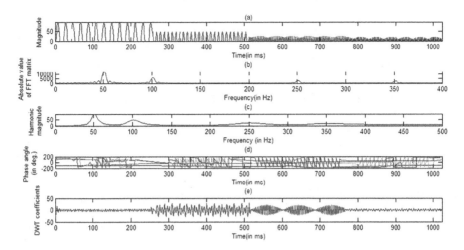

FIGURE 4.9 (a) Non-stationary signal, (b) frequency spectrum of the signal obtained from FFT, (c) frequency spectrum of the signal obtained from ST, (d) phase angle of the frequency components vs time, obtained from ST, and (e) magnitude of DWT coefficients at level 2 of the same signal.

TABLE 4.5

Magnitudes of Harmonic Components and Phase Angle Obtained from S-Transform

Frequency (in Hz)	Amplitude (in Volts)	Phase Angle (in degrees)
50	100	−110
100	50	−29.7
250	25	−29
350	20	61

Hence, DWT and ST are suitable in this case. Table 4.5 summarizes the amplitude of the harmonic components. Figure 4.9(e) clearly shows that the four frequency components occur at four different intervals of time.

Figure 4.9(d) shows that the phase angle for a particular frequency does not remain stationary with respect to the time for non-stationary signal, unlike the case of stationary signal as shown in Figure 4.8(d). Hence, it is difficult to ascertain the magnitude of the phase angle of a particular frequency component of a non-stationary signal from the S-matrix. Hence, in this case, the magnitude of phase angle of every frequency component can be accurately obtained from the argument of FFT matrix and is shown in Table 4.5

4.4 EFFECT OF NOISE

This section discusses the TFR analysis of noisy signals by FFT, ST, and DWT.

Gaussian white noise is added to the stationary and non-stationary signals shown in Figures 4.8(a) and 4.9(a). DWT, ST, and FFT have been employed on the noisy

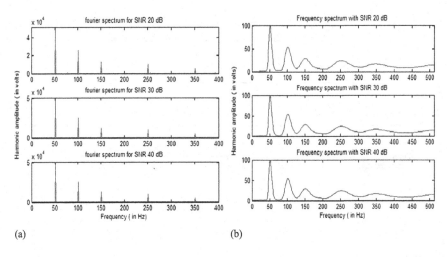

(a)   (b)

FIGURE 4.10   (a) Frequency spectrum of the noisy stationary signal obtained from FFT for different values of SNR and (b) Fourier spectrum of the noisy stationary signal obtained from ST for different values of SNR.

TABLE 4.6

Magnitudes of Harmonic Components and Their Corresponding Phase Angle of the Noisy Stationary Signal Obtained from ST and FFT for Different Values of SNR

	From ST				From FFT			
	Harmonic Amplitude (in Volt)				Phase Angle (in degree)			
Frequency (in Hz)	Pure Signal	SNR 20 dB	SNR 30 dB	SNR 40 dB	Pure Signal	SNR 20 dB	SNR 30 dB	SNR 40 dB
50	100	102	100.2	100.2	−60	−60	−60	−60
100	50	52	51	50	60	60	59.8	59.9
150	25	30	26.6	25.4	30	29	29.7	30
250	20	25.8	20.4	20.6	60	59	60	59.9
350	10	15.56	11.6	10.8	−40	−42	−40.7	−39.8

---

signals consisting of signal to noise ratio (SNR) of 20, 30, and 40 dB. The relevant outputs have been shown and discussed in the following sections.

4.4.1   Effect of Noise on a Pure Non-Sinusoidal Stationary Signal

The amplitude spectra in Figure 4.10 obtained both from FFT and ST remain almost unaffected by the presence of noise. The harmonic components are clearly visible and their magnitudes are summarized in Table 4.6. Distortion of the DWT coefficients in the presence of noise is clearly visible in Figure 4.11 (a). The profile of phase angle with respect to time, obtained from ST, gets distorted with the increase of the noise level in the signal (i.e., with the decrease of SNR value). The magnitudes of the phase angles for different values of SNR are provided in Table 4.6.

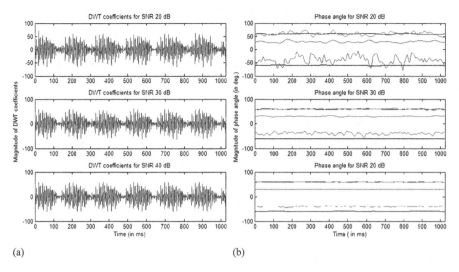

(a)   (b)

FIGURE 4.11 (a) Magnitude of DWT coefficients at level 2 of the noisy stationary signal for different values of SNR. (b) Phase angle of the frequency components of the noisy stationary signal vs time, obtained from ST, for different values of SNR.

4.4.2 Effect of Noise on a Pure Non-Stationary Signal

A noisy non-stationary signal has been studied in this section. The frequency spectrum using FFT and ST is shown in Figure 4.12. The magnitudes of the harmonic components and their corresponding phase angles are given in Table 4.7. It can be inferred from the observations made in this section that FFT and ST are more immune

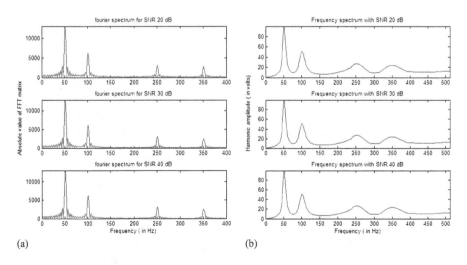

(a)   (b)

FIGURE 4.12 (a) Frequency spectrum of the noisy non-stationary signal obtained from FFT for different values of SNR. (b) Fourier spectrum of the noisy non-stationary signal obtained from ST for different values of SNR.

TABLE 4.7

Magnitudes of Harmonic Components and Their Corresponding Phase Angle of the Noisy Non-Stationary Signal Obtained from ST and FFT for Different Values of SNR

	From ST				From FFT			
	Harmonic Amplitude (in Volt)				Phase Angle (in degree)			
Frequency (in Hz)	Pure Signal	SNR 20 dB	SNR 30 dB	SNR 40 dB	Pure Signal	SNR 20 dB	SNR 30 dB	SNR 40 dB
50	100	99.7	99.7	99.7	−110	−110	−110.4	−110.4
100	50	50.9	50.9	50.9	−29.7	−30	−30	−29.6
250	25	26.6	26.6	26.6	−29	−28.6	−30.4	−29
350	20	22	22	22	61	63.4	60.6	60.8

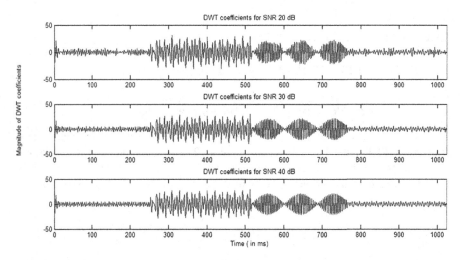

FIGURE 4.13 Magnitude of DWT coefficients at level 2 of the noisy non-stationary signal for different values of SNR.

to noise than DWT. Identification of harmonic components and determination of their magnitudes and phase angles can be accurately done with FFT and ST in the presence of noise. The profiles of the DWT coefficients for different values of SNRs are plotted in Figure 4.13

4.5   CONCLUSION

A comparative analysis of the three methods, i.e., FFT, DWT, and ST has been presented in this chapter by considering different signals simulated in MATLAB. Every method has its own utility and limitation with respect to the analysis of a particular signal. A summary of the capability of these tools of signal processing has been tabularized in Table 4.8.

TABLE 4.8
Comparison of FFT, DWT, and ST

	FFT	DWT	ST
TFR representation	No	Yes	Yes
Magnitude of frequency (in Hz)	Yes	No, because signal is decomposed into frequency bands at every level	Yes
Amplitude of frequency component	Cannot be obtained directly	No	Yes, directly from STA matrix
Phase angle of frequency component	Yes	No	Yes
Frequency resolution	Highest	Minimum	Medium
Phase resolution	Highest, for both stationary and non-stationary signals	Zero	High, for stationary signal but low, for non-stationary signal
Immunity to noise	Highest, in both frequency and phase resolution	Minimum	High in frequency resolution and low in phase resolution

It is clear from the above discussions that the combination of FFT and ST gives accurate information of the phase and amplitude of the harmonic components of any real-time signal. DWT decomposes the signal into several bands of frequencies at different levels of decomposition, and it is computationally fast for a large scale of data. The signal amplitude and its phase angle cannot be obtained directly from DWT.

REFERENCES

1. B. Boashash, "Notes on the use of the wigner distribution for time frequency signal analysis", *IEEE Transactions on Acoustics Speech and Signal Processing*, vol. 26, no. 9, 1987.
2. F. Hlawatsch and G. F. Boudreuax-Bartels, "Linear and quadratic time frequency signal representations", *IEEE SP Magazine*, pp. 21–67, April 1992.
3. N. E. Sejdić, I. Djurović, and J. Jiang, "Time-frequency feature representation using energy concentration: An overview of recent advances", *Digital Signal Processing*, vol. 19, no. 1, pp. 153–183, January 2009.
4. R. N. Bracewell, *The Fourier Transform and Its Applications*, McGrawHill Book Company, New York, 1978.
5. E. O. Brigham, *The Fast Fourier Transform*, Prentice-Hall Inc., Englewood Cliffs, New Jersey, 1974.
6. R. G. Stockwell, L. Mansinha, and R. P. Lowe, "Localization of the complex spectrum: The S transform", *IEEE Transactions on Signal Processing*, vol. 44, no. 4, pp. 998–1001, 1996.

7. R. G. Stockwell, S-transform analysis of gravity wave activity from a small scale network of airglow imagers, PhD thesis, University of Western Ontario, London, Ontario, Canada, 1999.

8. R. A. Brown and R. Frayne, *A fast discrete S-transform for biomedical signal processing, 2008 30th Annual International Conference of the IEEE Engineering in Medicine and Biology Society*, Vancouver, BC, Canada, 2008, pp. 2586–2589, doi: 10.1109/IEMBS.2008.4649729.

9. I. Daubechies, "The wavelet transform, time-frequency localization and signal analysis", *IEEE Transactions on Information Theory*, vol. 36, no. 5, Sep 1990.

10. O. Rioul and M. Vetterli, "Wavelets and signal processing", *IEEE SP Magazine*, vol. 8, pp. 14–38, 1991.

11. M. Farge, "Wavelet transforms and their application to turbulence", *Annual Review of Fluid Mechanics*, vol. 24, pp. 395–457, 1992.

12. R. K. Young, *Wavelet Theory and its Applications*, Kluwer Academic Publishers, Dordrecht, 1993.

13. J. Ding, *Time-Frequency Analysis and Wavelet Transform Course Note*, The Department of Electrical Engineering, National Taiwan University (NTU), Taipei, Taiwan, 2007.

5 Neural Network

5.1 INTRODUCTION

An Artificial Neural Network (ANN) can be described as a set of elementary neurons that are usually connected in biologically inspired architectures and organized in several layers [1–2]. The structure of a feed-forward ANN, also called as the perceptron, is shown in Figure 5.1. There are Ni numbers of neurons in each ith layer, and the inputs to these neurons are connected to the previous layer neurons. The input layer is fed with the excitation signals. An elementary neuron is like a processor that produces an output by performing a simple non-linear operation on its inputs [3]. A weight is attached to each and every neuron, and training an ANN is the process of adjusting different weights tailored to the training set. An ANN learns to produce a response based on the inputs given by adjusting the node weights. Hence, we need a set of data referred to as the training data set, which is used to train the neural network.

Due to their outstanding pattern recognition abilities, ANNs are used for several purposes in a wide variety of fields including signal processing, computers, and decision-making. Some important notes on ANNs are [4]:

- Either signal features are extracted using certain measuring algorithms or even unprocessed samples of the input signals are fed into the ANN.
- The most recent along with a few older samples of the signals are fed into the ANN.

The output provided by the neural network corresponds to the concerned decision which might be the type of fault, existence of a fault, or the location of a fault.

- The most important factor that affects the functionality of the ANN is the training pattern that is employed for the same.
- Pre-processing and post-processing techniques may be employed as well to enhance the learning process and reduce the training time of the ANN.

One of the biggest drawbacks of applications that make use of ANNs is that no well-defined guide is available for selecting the ideal number of hidden layers to be used and the number of neurons per each hidden layer [5]. A vital feature of ANN is its dedication to parallel computing. Hence it can produce a correct output corresponding to any input even if the concerned input was not fed into the ANN during the training process. Another challenge in the ANN-based application development was to synthesize the algorithm for the adaptive learning process. The back-propagation algorithm is the basic algorithm in which the neuron weights are adjusted in consecutive steps to minimize the error between the actual and the desired outputs. This process is known as supervised learning.

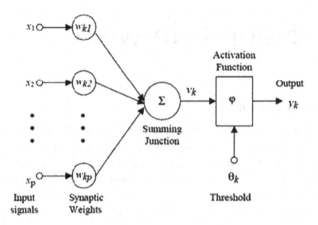

FIGURE 5.1   Model of an ANN.

5.2   MATHEMATICAL MODEL OF A NEURON

A neuron is an information processing unit that is fundamental to the operation of a neural network. The three basic elements of the neuron model are:

1. A set of weights, each of which is characterized by a strength of its own. A signal x_j connected to a neuron k is multiplied by the weight w_{kj}. The weight of an artificial neuron may lie in a range that includes negative as well as positive values.
2. An adder for summing the input signals, weighted by the respective weights of the neuron.
3. An activation function for limiting the amplitude of the output of a neuron. It is also referred to as a squashing function which squashes the amplitude range of the output signal to some finite value.

5.3   NETWORK ARCHITECTURES

There are three fundamental different classes of network architectures [6]:

1. Single-layer Feedforward Networks
   In a layered neural network, the neurons are organized in the form of layers. In the simplest form of a layered network, we have an input layer of source nodes that projects onto an output layer of neurons, but not vice versa. This network is strictly a Feedforward type. In a single-layer network, there is only one input and one output layer. Input layer is not counted as a layer since no mathematical calculations take place at this layer (Figure 5.2).
2. Multilayer Feedforward Networks
   The second class of a Feedforward neural network distinguishes itself by the presence of one or more hidden layers, whose computational nodes are correspondingly called hidden neurons. The function of a hidden neuron is to intervene between the external input and the network output in some useful manner.

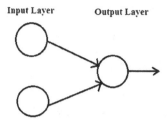

FIGURE 5.2 Single-layer Feedforward Network.

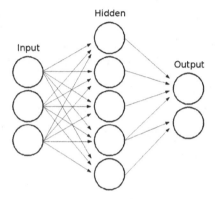

FIGURE 5.3 Multilayer Feedforward Network.

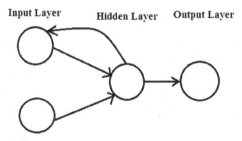

FIGURE 5.4 Recurrent Network.

By adding more hidden layers, the network is enabled to extract higher order statistics. The input signal is applied to the neurons in the second layer. The output signal of second layer is used as input to the third layer, and so on for the rest of the network (Figure 5.3).

3. Recurrent Networks

A recurrent neural network has at least one feedback loop. A recurrent network may consist of a single layer of neurons with each neuron feeding its output signal back to the inputs of all the other neurons. Self-feedback refers to a situation where the output of a neuron is fed back into its own input. The presence of feedback loops has a profound impact on the learning capability of the network and on its performance (Figure 5.4).

5.4 LEARNING PROCESSES

Learning rule [7] signifies a procedure for modifying the weights and biases of a network. The purpose of learning rule is to train the network to perform some tasks, and it has three categories as given below:

1. Supervised learning
   The learning rule is provided with a set of training data of proper network behavior. As the inputs are applied to the network, the network outputs are compared to the targets. The learning rule is then used to adjust the weights and biases of the network in order to move the network outputs closer to the targets.
2. Reinforcement learning
   It is similar to supervised learning, except that, instead of being provided with the correct output for each network input, the algorithm is only given a grade. The grade is a measure of the network performance over some sequence of inputs.
3. Unsupervised learning
   The weights and biases are modified in response to network inputs only. There are no target outputs available. Most of these algorithms perform some kind of clustering operation. They learn to categorize the input patterns into a finite number of classes.

5.5 BACK PROPAGATION ALGORITHM

Multiple layer perceptrons have been applied successfully to solve some difficult diverse problems by training them in a supervised manner with a highly popular algorithm known as the error back-propagation algorithm. This algorithm is based on the error–correction learning rule. It may be viewed as a generalization of an equally popular adaptive filtering algorithm – the least mean square (LMS) algorithm. Error back-propagation learning consists of two passes through the different layers of the network: a forward pass and a backward pass. In the forward pass, an input vector is applied to the nodes of the network, and its effect propagates through the network layer by layer. Finally, a set of outputs is produced as the actual response of the network. During the forward pass, the weights of the networks are all fixed. During the backward pass, the weights are all adjusted in accordance with an error correction rule. The actual response of the network is subtracted from a desired response to produce an error signal. This error signal is then propagated backward through the network, against the direction of synaptic connections. The weights are adjusted to make the actual response of the network move closer to the desired response.

A multilayer perceptron has three distinctive characteristics:

- The model of each neuron in the network includes a non-linear activation function.
  The three commonly used functions are shown in Figure 5.5.

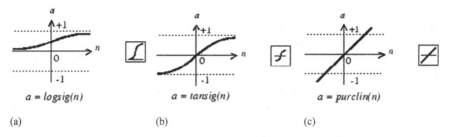

$$a = logsig(n) \qquad\qquad a = tansig(n) \qquad\qquad a = purelin(n)$$

(a)   (b)   (c)

FIGURE 5.5 Commonly used transfer functions for multilayer networks (a) Log-Sigmoid Transfer Function, (b) Tan-Sigmoid Transfer Function, (c) Linear Transfer Function.

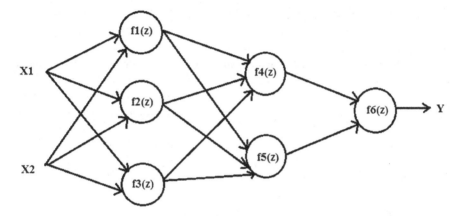

FIGURE 5.6 Three-layer neural network with two inputs and single output.

Sigmoid output neurons are often used for pattern recognition problems, while linear output neurons are used for function fitting problems [8].

- The network contains one or more layers of hidden neurons which enable the network to learn complex tasks.
- The network exhibits a high degree of connectivity. A change in the connectivity of the network requires a change in the population of their weights.

5.5.1 LEARNING PROCESS

The learning process can be illustrated by a three-layer neural network [7] with two inputs and one output, as shown in Figure 5.6.

Signal z is adder output signal, and $y = f(z)$ is the output signal of non-linear element. Signal y is also the output signal of neuron. The training data set consists of input signals ($x1$ and $x2$) assigned with corresponding target (desired output) Y. The network training is an iterative process. The network can be trained for function approximation (non-linear regression), pattern association, or pattern classification. The training process requires a set of examples of proper network behavior – network inputs p and target outputs t. The default performance function for feedforward

Fundamentals networks is mean square error (mse) – the average squared error between the network outputs and the target outputs. There are several different training algorithms for feedforward networks. All of these algorithms use the gradient of the performance function to determine how to adjust the weights to minimize performance. The gradient is determined using a technique called back-propagation, which involves performing computations backward through the network.

It is very difficult to know which training algorithm will be the fastest for a given problem [8]. It will depend on many factors, including the complexity of the problem, the number of data points in the training set, the number of weights and biases in the network, the error goal, and whether the network is being used for pattern recognition (discriminant analysis) or function approximation (regression).

The nodes in the neural network are divided into a number of layers, the input layer, one or more hidden layers, and the output layer. The required number of hidden layers and the number of nodes for each layer are problem oriented. The neural network with one hidden layer is called a three-layer neural network [9].

In the feedforward network structure, the signals propagate only from the input layer to the hidden layer and from the hidden to output layer. The connections between the nodes within the same layer or from the input layer directly to the output layer are not allowed. The nodes in the hidden and output layers act as neurons doing activation and output functions.

For each node in the input layer, the output is the same as that of the input. If net_i is the net input to a neuron i, then output of neuron i is given by

$$o_i = net_i \tag{5.1}$$

The net input to each neuron j in the hidden layer is given by

$$net_j = \sum_{i=1}^{NI} w_{ji} o_i, \quad j = 1,\ldots\ldots\ldots NJ \tag{5.2}$$

where NI and NJ are the number of nodes in the input layer and hidden layer, respectively.

The output of neuron j is a sigmoidal activation function of input and is given by

$$o_j = \frac{1}{1 + e^{-(net_j + \theta_j)}} \tag{5.3}$$

$$= f_j\left(net_j, \theta_j\right) \tag{5.4}$$

where Θ_j is a threshold or bias. Θ_j can be regarded as connection weight between node j and a fictitious node in the previous layer, and the output value of which is always unity.

Similarly, for a neuron in the output layer k, the net input is given by,

$$net_k = \sum_{j=1}^{NJ} w_{kj} o_j, \quad k = 1,\ldots\ldots\ldots, NK \tag{5.5}$$

where NK is the number of nodes in the output layer. The corresponding output is given by,

$$o_k = \frac{1}{1 + e^{-(net_k + \theta_k)}} \tag{5.6}$$

$$= f_k(net_k, \theta_k) \tag{5.7}$$

In the learning or training of such a net, the pattern is presented as input and the set of weights in all the connecting links and also all the thresholds in the nodes are adjusted in such a way that the desired outputs are obtained at the output nodes. Once this adjustment has been achieved by the net, another pair of input–output pattern is presented to the net and the net is required to learn that association also. In fact, the net is adjusted to find a single set of weights and biases that will satisfy all the (input–output) pairs presented to it.

During learning or training process, the output will not be the same as the desired values. For each pattern p, the sum of the squared errors to be minimized is given by

$$E_p = \frac{1}{2} \sum_{k=1}^{NK} (t_{pk} - o_{pk})^2 \tag{5.8}$$

where t_{pk} and o_{pk} are the target output value and computed output value at output node k, respectively.

The overall measure of the error for NP input patterns is given by,

$$E = \sum_{p=1}^{NP} E_p \tag{5.9}$$

In the generalized delta rule formulated by Rumelhart, Hinton, and Williams [9] for learning the weights and biases, the procedure for learning the correct set of weights is to vary the weights in a manner calculated to reduce the error E_p as rapidly as possible. The corrections to the weights are made sequentially on the basis of the learning to be carried out for the sequence of the patterns one at a time.

Omitting the subscript p for convenience, Equation (5.8) can be written as,

$$E = \frac{1}{2} \sum_{k=1}^{NK} (t_k - o_k)^2 \tag{5.10}$$

Convergence toward improved values for the weights and thresholds are obtained by taking incremental changes Δw_{kj} proportional to $-\dfrac{\partial E}{\partial w_{kj}}$, given by,

$$\Delta w_{kj} = -\eta \frac{\partial E}{\partial w_{kj}} \tag{5.11}$$

where η is the learning rate. The partial derivative $\dfrac{\partial E}{\partial w_{kj}}$ can be evaluated by the chain rule [10], given by,

$$\frac{\partial E}{\partial w_{kj}} = \frac{\partial E}{\partial net_k} \cdot \frac{\partial net_k}{\partial w_{kj}} \qquad (5.12)$$

Using Equation (5.5),

$$\frac{\partial net_k}{\partial w_{kj}} = \frac{\partial}{\partial w_{kj}} \sum w_{kj} o_j = o_j \qquad (5.13)$$

Let,

$$\partial_k = -\frac{\partial E}{\partial net_k} \qquad (5.14)$$

Therefore,

$$\Delta w_{kj} = \eta \partial_k . o_j \qquad (5.15)$$

To compute $\partial_k = -\dfrac{\partial E}{\partial net_k}$, the chain rule is used [10] to express the partial derivative in terms of two factors, one expressing the rate of change of error with respect to the output o_k, and the other expressing the rate of change of the output of the node k with respect to the input to that same node.

Hence,

$$\partial_k = -\frac{\partial E}{\partial net_k} = -\frac{\partial E}{\partial o_k} \cdot \frac{\partial o_k}{\partial net_k} \qquad (5.16)$$

From Equation (5.10),

$$\frac{\partial E}{\partial o_k} = -\left(t_k - o_k\right) \qquad (5.17)$$

Again from Equation (5.6)

$$\begin{aligned}
\frac{\partial o_k}{\partial net_k} &= \frac{\partial}{\partial net_k}\left(\frac{1}{1+e^{-(net_k+\theta_k)}}\right) \\
&= \frac{e^{-(net_k+\theta_k)}}{\left(1+e^{-(net_k+\theta_k)}\right)^2} \\
&= \frac{1}{\left(1+e^{-(net_k+\theta_k)}\right)}\left(1-\frac{1}{1+e^{-(net_k+\theta_k)}}\right) \\
&= o_k\left(1-o_k\right)
\end{aligned} \qquad (5.18)$$

Therefore,

$$\delta_k = (t_k - o_k) o_k (1 - o_k) \tag{5.19}$$

Now, for the lower layer where weights do not affect output nodes directly, it can be written as [10],

$$\Delta w_{ji} = -\eta \frac{\partial E}{\partial w_{ji}} \tag{5.20}$$

$$= -\eta \frac{\partial E}{\partial net_j} \cdot \frac{\partial net_j}{\partial w_{ji}}$$

$$= -\eta \frac{\partial E}{\partial net_j} o_i \tag{5.21}$$

$$= \eta \delta_j o_i$$

where

$$\delta_j = -\frac{\partial E}{\partial net_j}, \tag{5.22}$$

or,

$$\delta_j = -\frac{\partial E}{\partial o_j} \frac{\partial o_j}{\partial net_j} \tag{5.23}$$

similar to Equation (5.18)

$$\frac{\partial o_j}{\partial net_j} = o_j (1 - o_j) \tag{5.24}$$

However, the factor $\frac{\partial E}{\partial o_j}$ cannot be evaluated directly. Instead, it can be represented by the quantities that are known and other quantities that can be evaluated. Hence,

$$\frac{\partial E}{\partial o_j} = \sum_{k=1}^{NK} \frac{\partial E}{\partial net_k} \frac{\partial net_k}{\partial o_j} \tag{5.25}$$

or,

$$= -\eta \frac{\partial E}{\partial net_j} \cdot \frac{\partial net_j}{\partial w_{ji}}$$

$$= -\eta \frac{\partial E}{\partial net_j} o_i \tag{5.26}$$

$$= \eta \delta_j o_i$$

Hence, from Equations (5.23–5.26)

$$\delta_j = o_j \left(1 - o_j\right) \sum_{k=1}^{NK} \delta_k \cdot w_{kj} \tag{5.27}$$

The deltas at an internal node can be evaluated in terms of the deltas at an upper layer. Thus, starting at the highest layer (upper layer), δ_ks are evaluated using Equation (5.19), and the errors can then propagate backward to lower layers.

Summarizing and using the subscript p to denote the pattern number,

$$\Delta w_{kj}\left(p\right) = \eta \delta_{pk} \cdot o_{pj} \tag{5.28}$$

and

$$\delta_{pk} = \left(t_{pk} - o_{pk}\right) o_{pk} \left(1 - o_{pk}\right) \tag{5.29}$$

for the output layer nodes.

Similarly, for hidden layer nodes

$$\Delta w_{ji}\left(p\right) = \eta \delta_{pj} \cdot o_{pi} \tag{5.30}$$

$$\delta_{pj} = o_{pj} \left(1 - o_{pj}\right) \sum_{k=1}^{NK} \delta_{pk} w_{kj} \tag{5.31}$$

It is noted that the threshold θ_j is learned in the same manner as are the other weights. The learning pattern therefore consists of the net starting off with a random set of weight values, choosing one of the training set patterns and using that pattern as the input, evaluating the output in a feedforward manner. The errors at the outputs are initially quite large, which necessitates changes in the weights. Using the back-propagation procedure, the net calculates Δw_{ji} (p), and correction is made for all w_{ji} for that particular pattern p. This procedure is repeated for all the patterns in the training set, and the first iteration is thus complete. Patterns are again presented in the subsequent iteration, and the weights w_{ji} are corrected at every presentation of patterns. In a successful learning process, the system error will decrease with the number of iterations, and the procedure will converge to a stable set of weights which will exhibit only small fluctuations in value as further learning is attempted.

5.5.2 TRAINING ALGORITHMS

Back-propagation can train multilayer feed-forward networks with differentiable transfer functions to perform function approximation, pattern association, and pattern classification. (Other types of networks can be trained as well, although the multilayer network is most commonly used.) The term back-propagation refers to the process by which derivatives of network error, with respect to network weights and biases, can be computed. This process can be used with a number of different optimization strategies. The architecture of a multilayer network is not completely constrained by the problem to be solved. The number of inputs to the network is constrained by the problem, and the number of neurons in the output layer is constrained by the number of outputs required by the problem. However, the number of layers between network inputs and the output layer and the sizes of the layers are up to the designer. The two-layer sigmoid/linear network can represent any functional relationship between inputs and outputs if the sigmoid layer has enough neurons. There are several different back-propagation training algorithms and a few of them have been enlisted below:

- Basic gradient descent
- Gradient descent with momentum
- Resilient back-propagation
- Adaptive learning rate
- Levenberg–Marquardt algorithm (LM)
- Bayesian regularization

They have a variety of different computation and storage requirements, and no one algorithm is best suited to all locations. However, LM algorithm is the fastest training algorithm [8] for networks of moderate size. It has memory reduction feature for use when the training set is large. The present work in the thesis involves LM algorithm for the purpose of training.

5.6   PROBABILISTIC NEURAL NETWORK

Probabilistic neural networks (PNNs) can be used for classification problems. When an input is presented, the first layer computes distances from the input vector to the training input vectors, and produces a vector whose elements indicate how close the input is to a training input. The second layer sums these contributions for each class of inputs to produce as its net output a vector of probabilities. Finally, a *compete* transfer function on the output of the second layer picks the maximum of these probabilities, and produces a one for that class and a zero for the other classes.

The PNN model is one among the supervised learning networks and has the following features distinct from those of other networks in the learning processes [11–12].

- It is implemented using the probabilistic model, such as Bayesian classifiers. A PNN is guaranteed to converge to a Bayesian classifier provided that it is given enough training data.
- No learning processes are required.
- No need to set the initial weights of the network.
- No relationship between learning processes and recalling processes.
- The difference between the inference vector and the target vector is not used to modify the weights of the network.

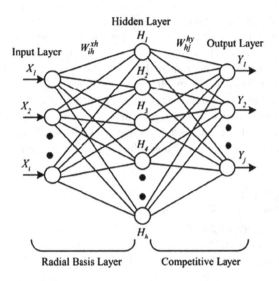

FIGURE 5.7   Architecture of a PNN.

The learning speed of the PNN model is very fast making it suitable in real time for fault diagnosis and signal classification problems. Figure 5.6 shows the architecture of a PNN model that is composed of the radial basis layer and the competitive layer. In the signal-classification application, the training examples are classified according to their distribution values of probabilistic density function (pdf), which is the basic principle of the PNN. A simple pdf is as follows (Figure 5.7):

$$f_k(X) = \frac{1}{N_k} \sum_{j=1}^{N_k} \exp\left(-\frac{X - X_{kj}}{2\sigma^2}\right) \tag{5.32}$$

Modifying and applying (5.1) to the output vector H of the hidden layer in the PNN is as follows:

$$H_h = \exp\left(\frac{-\sum_i \left(X_i - W_{ih}^{xh}\right)^2}{2\sigma^2}\right) \tag{5.33}$$

$$net_j = \frac{1}{N_j} \sum_h W_{hj}^{hy} H_h \quad \text{and} \quad N_j = \sum_h W_{hj}^{hy}$$

$$net_j = \max_k(net_k) \text{ then } y_j = 1, \text{ else } y_j = 0,$$

where

i = number of input layers;
h = number of hidden layers;
j = number of output layers;

k = number of training examples;

N = number of classifications (clusters);

σ = smoothing parameter (standard deviation);

X = input vector;

and where $\|X - X_{kj}\|$ is the Euclidean distance between the vectors X and X_{kj}, i.e., $\|X - X_{kj}\| = \sum_i (X - X_{kj})^2$; W_{ih}^{xh} is the connection weight between the input layer X and the hidden layer H; and W_{hj}^{hy} is the connection weight between the hidden layer H and the output layer Y.

PNNs can be used for classification problems [8]. Their design is straightforward and does not depend on training. A PNN is guaranteed to converge to a Bayesian classifier provided it is given enough training data. These networks generalize well.

5.7 CONCLUSION

It is very essential to investigate and analyze the advantages of a particular neural network structure and learning algorithm before choosing it for an application because there should be a trade-off between the training characteristics and the performance factors of any neural network. Neural Network is a reliable and attractive scheme for fault classification and estimation of fault location in power system networks. The back-propagation algorithm is the basic algorithm in which the neuron weights are adjusted in consecutive steps to minimize the error between the actual and the desired outputs. A PNN using the probability density function and Bayes decision rule is efficient in data classification. In the present thesis, PNN has been used for identifying the type of fault, faulty phase(s), and fault zone in a power system network. Once the fault is identified; BPNN has been implemented for obtaining the particular location of fault in the network.

REFERENCES

1. A. Cichoki and R. Unbehauen, *Neural Networks for Optimization and Signal Processing*, John Wiley & Sons, Inc., New York, 1993.
2. Suhaas Bhargava Ayyagari, Artificial neural network based fault location for transmission lines, Thesis, 2011, University of Kentucky.
3. S. Haykin, *Neural Networks, A Comprehensive Foundation*, Macmillan Collage Publishing Company, Inc., New York, 1994.
4. M. Kezunovic, "A survey of neural net applications to protective relaying and fault analysis", *International Journal of Engineering Intelligent Systems for Electronics, Engineering and Communications*, vol. 5, no. 4, pp. 185–192, 1997.
5. David Kriesel, A brief introduction to neural networks. http://www.dkriesel.com/en/science/neural_networks.
6. P. Werbos, "Generalization of backpropagation with application to recurrent gas market model", *Neural Networks*, vol. 1, pp. 339–356, 1988.
7. An introduction to neural networks. www2.econ.iastate.edu/tesfatsi/NeuralNetworks.CheungCannonNotes.pdf.

8. Howard Demuth, Mark Beale, and Martin Hagan, The MathWorks user's guide for MATLAB and Simulink, Neural Network Toolbox User's Guide, Version 6.

9. D. E. Rumelhart, G. E. Hinton and R. J. Williams, "Learning internal representation by error propagation", *Parrallel Distributed Processing*, vol. 1, pp. 318–362, Cambridge, MA, MIT Press, 1986.

10. Dipak Ray, Studies on reliability and static security of power system, Thesis, 1993.

11. S. Mishra, C. N. Bhende, and B. K. Panigrahi, "Detection and classification of power quality disturbances using S-transform and probabilistic neural network", *IEEE Transactions on Power Delivery*, vol. 23, no. 1, pp. 280–287, Jan 2008.

12. Maryam Mirzaei, Mohd Zainal Abidin Ab. Kadir, Hashim Hizam, and Ehsan Moazami, "Comparative analysis of probabilistic neural network, radial basis function, and feed-forward neural network for fault classification in power distribution systems", *Electric Power Components and Systems*, vol. 39, no. 16, pp. 1858–1871, 2011.

6 Fault Analysis in Single-Circuit Transmission Line Using S-Transform and Neural Network

6.1 INTRODUCTION

This chapter demonstrates a technique for diagnosis of fault type and faulty phase on overhead transmission lines. A method for computation of fault location is also incorporated in this work. The proposed method is based on the multiresolution S-transform, which is used for generating complex *S*-matrices of the current signals measured at the sending and receiving ends of the line. The peak magnitude of the absolute value of every *S*-matrix is noted. The phase angle corresponding to every peak component is obtained from the argument of the relevant *S*-matrix. These features are used as input vectors of a probabilistic neural network for fault detection and classification. Detection of faulty phase(s) is followed by estimation of fault location. The voltage signal of the affected phase is processed to generate the *S*-matrix. The frequency components of the *S*-matrices for different fault locations are used as input vectors for training a back-propagation neural network (BPNN). The results are obtained with satisfactory accuracy and speed. All the simulations have been done in MATLAB (The MathWorks, Natick, Massachusetts, USA) environment for different values of fault locations, fault resistances, and fault inception angles. The effect of noise on both the current and voltage signals has been investigated.

The present chapter is organized as follows. The features of classification have been described in Section 6.2. The simulated power system network is discussed in Section 6.3. Section 6.4 explains the PNN network in detail. The results of simulation and classification have been tabulated in Section 6.5. The effect of noise in the suggested technique is studied in Section 6.6. Section 6.7 describes the BPNN architecture for estimation of fault location. The conclusion of the present work is presented in Section 6.8. A summary of future work to be implemented is given in Section 6.9.

6.2 FEATURE EXTRACTION BY S-TRANSFORM

The simulated signals in the present work are in discrete form, and therefore the discrete version of ST has been employed here for signal processing. The output of discrete ST (DST) is a complex matrix (S-matrix) where the row corresponds to the

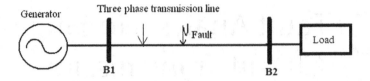

FIGURE 6.1   Single Line diagram of three-phase network.

frequency and the column pertains to the time. The steps of computation of DST have been elucidated in [1–3], and the original software code developed by Stockwell [4] in Matlab has been employed here with modifications for feature extraction. The magnitudes of the harmonic components can be obtained from the absolute value of the S-matrix (STA). The argument of the S-matrix (STP) gives the phasor components of the harmonic elements.

The steps of obtaining classification features have been listed below:

- The absolute matrices STA1 and STA2 are obtained for the current signals of B1 (sending end of the network shown in Figure 6.1) and B2 (receiving end of the network shown in Figure 6.1).
- The STP1 and STP2 for the same S-matrices are also evaluated.
- Maximum values (S1 and S2) of the two STA matrices are obtained.
- Suppose $S1 \angle \phi_1$ is an element of the S-matrix corresponding to current signal of B1, where S1 is the maximum value of the STA1 matrix and ϕ_1 is the corresponding phasor component. Similarly, $S2 \angle \phi_2$ is obtained corresponding to current signal of B2.
- Ratio, $Q = \dfrac{S1}{S2}$ and phase difference, $P = \phi_1 - \phi_2$ are evaluated.

It has been seen from all the simulations that Q is close to unity under no fault condition for all the three phases. During any fault this Q is greater than unity for the faulty phase(s). Also, the magnitude of P under normal condition is almost the same for all the three phases. Otherwise, P is different for all the phases and it is less than the normal value for the short-circuited phase(s). Section 6.7 describes the features required for computation of fault location.

6.3   POWER SYSTEM UNDER STUDY

Expert techniques of fault classification in single-circuit transmission line are demonstrated in [5–7] based on wavelet transform and decision-tree methodology, respectively. An application of the discrete wavelet transform (DWT) and BPNNs is proposed in [5] for fault diagnosis on single-circuit transmission line. Wavelet entropy principle combined with PNN is applied for fault classification in [6]. However, these techniques do not include the effect of noise on the current/voltage signals.

TABLE 6.1

Line and System Parameters Used for Generation of Training and Testing Patterns

Parameters	Training Data Set	Testing Data Set
Fault types	ten types of faults	ten types of faults
Fault inception angle (θ, in deg)	0 and 90	0 and 45
Fault resistance (R_F, Ω)	0, 40, 80, and 100	0, 20, and 60
Fault location from sending end (D, in km)	10, 30, 50, 70, 90, 110, 130, 150, 170, 190, 210, 230, 250, 270, and 290	20, 40, 60, 80, 100, 120, 140, 160, 180, 200, 220, 240, 260, and 280
Total number of patterns	10×2×4×15 = 1200	10×2×3×14 = 840

A 400 kV, 50 Hz, three-phase power system network is simulated using the Simpower Toolbox of MATLAB-7 and is shown in Figure 6.1. The length of the transmission line is 300 km. A three-phase balanced load is connected at the receiving end (B2) of the transmission line. The parameters of the balanced load are given below:

- Load Impedance: $(720+j11)\Omega$
- P.f: 0.9
- MVA rating: 200

The sampling time of all the signals is taken to be 78.28 μs, and the time period of simulation in MATLAB has been taken up to 0.04 secs. The sampling frequency is 12.8 kHz. The following ten types of faults are simulated in this system:

- Single Line-Ground fault for phase A, B, and C, respectively (i.e., AG, BG, and CG).
- Double Line fault (i.e., AB, BC, and CA).
- Double Line-Ground fault (i.e., ABG, BCG, and CAG).
- Three-phase fault, i.e., LLL.

All the faults have been initiated at 29 different locations starting from B1, each being 10 km apart. The simulation parameters used for data generation are summarized in Table 6.1.

6.4 PNN-BASED FAULT CLASSIFICATION

The structure of the PNN used for fault classification consists of two hidden layers. The first hidden layer is based on the radial basis transfer function, and the second hidden layer is based on the competitive transfer function. The size of the input layer for the PNN is 6×11, where 6 represents the input features (i.e., QA, PA, QB, PB,

QC, and PC of the three phases) and 11 is the number of fault types, including the no-fault condition. The key advantages of the PNN are its fast-training process, its inherently parallel structure, which is guaranteed to converge to an optimal classifier as the size of the representative training set increases; and its ability to add or remove training samples without extensive retraining. The normal signal and different faulty signals have been categorized as AG (1), BG (2), CG (3),AB (4), BC (5), CA (6), ABG (7), BCG(8), CAG (9),LLL (10), and normal (11), respectively. The size of the output layer of the PNN is 1×11. Since a particular type of fault is likely to occur at one location at a given point of time, the output of the PNN is the category of that particular fault, i.e., one numeric integer ranging from one to 11. The features of ten faulty voltage signals for each type of fault have been used for training. As the training data set consists of 1200 elements, the PNN network has been rigorously tested with 840 data signals by changing the training samples 12 times.

6.5   RESULTS OF SIMULATION AND PNN CLASSIFIER

The voltage profiles for AG, AB, ABG, and LLL type of faults are shown in Figure 6.2

Figures 6.2(a)–(d) show that in the case of phase–ground and phase–phase short circuit faults, there are very fast high-frequency oscillations in the voltage profile of all the three phases A, B, and C at the onset of the short between any phase(s) to ground (excepting in phase C for AB fault). The amplitude of the shorted phases at the instant of fault occurrence, however, depends on the time of initiation of the fault. In this article, the time of occurrence of the fault as mentioned before is taken at three different instants, i.e., after 10 ms (i.e., after one half cycle), 12.5 ms, and 15 ms. The amplitudes of sending end phase voltages of A, B, and C after 10 ms are 129.26 kV, 48.80 kV, and −178.06 kV, respectively. Table 6.2 shows the values of classification features of all the fault types that have occurred at distances of 10 km and 200 km from the sending end of the transmission line (B1).

The results of the PNN classifier are shown in Table 6.3. It is noticed from Table 6.3 that the total number of misclassified cases is three. The average accuracy of classification in the present study is 99.6%.

6.6   EFFECT OF NOISE ON FAULT DIAGNOSIS

In order to analyze the performance of the classification program with noisy input signals, simulations were performed with white gaussian noise added to the simulated voltage and current signals by considering a noise level of 20 dB SNR. The entire testing data consisting of 840 data have been impregnated with this noise by simulation in MATLAB. The noisy voltage profiles in the case of AG, AB, ABG, and LLL faults are shown in Figure 6.3.

As an illustration, Table 6.4 shows the values of classification features for a particular fault condition in noisy environment.

Table 6.5 summarizes the results for fault location computation with 20 dB noise added to the voltage signals.

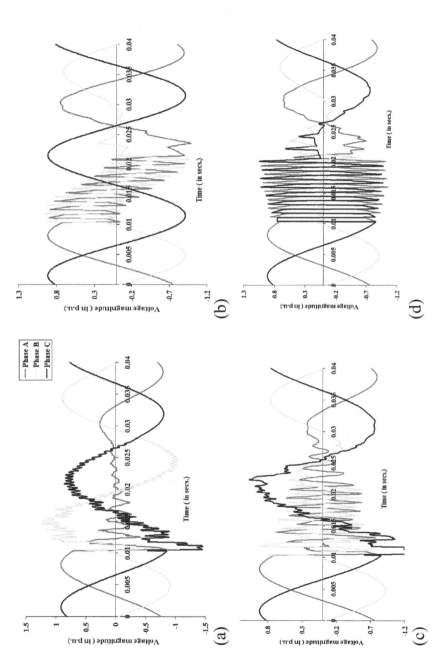

FIGURE 6.2 Voltage profiles of the three phases for (a) AG, (b) AB, (c) ABG, and (d) LLL type of faults at 100 km from B1, $R_F = 0$, and $\theta = 0^0$.

TABLE 6.2

Magnitudes of Classification Features for Different Types of Fault

D=10 km, $R_F = 0\ \Omega$, $\theta = 0^0$	Q A	P A	Q B	P B	Q C	P C
PURE	1.01	57.75^0	1.01	57.89^0	1.01	57.75^0
AG	1.88	20.32^0	0.97	59.02^0	1.02	55.16^0
BG	1.02	55.39^0	1.94	20.46^0	0.97	58.44^0
CG	0.97	58.26^0	1.02	55.92^0	1.56	28.33^0
AB	1.85	55.52^0	1.05	10.21^0	1.01	57.73^0
BC	1.01	57.73^0	1.87	57.35^0	1.11	8.87^0
CA	1.17	33.75^0	1.01	57.89^0	1.44	57.87^0
ABG	1.84	26.53^0	2.20	24.31^0	0.96	55.71^0
BCG	0.97	56.13^0	1.88	25.67^0	1.63	25.91^0
CAG	2.02	19.31^0	0.97	56.94^0	1.64	32.15^0
ABC	2.12	32.02^0	1.77	17.98^0	1.69	29.79^0
D=200 km, $R_F = 10\ \Omega$, $\theta = 45^0$	Q A	P A	Q B	P B	Q C	P C
PURE	1.00	57.77^0	1.01	57.86^0	1.01	57.71^0
AG	1.86	30.94^0	0.95	58.37^0	1.03	54.90^0
BG	1.03	55.01^0	1.78	32.55^0	0.95	57.84^0
CG	0.98	57.66^0	1.02	56.69^0	1.25	49.51^0
AB	1.78	57.28^0	1.08	12.95^0	1.01	57.72^0
BC	1.01	57.74^0	1.48	56.65^0	1.12	30.27^0
CA	1.42	44.56^0	1.01	57.88^0	1.32	66.35^0
ABG	1.90	30.99^0	1.72	26.98^0	0.97	55.21^0
BCG	1.01	55.16^0	1.89	47.49^0	1.33	46.62^0
CAG	1.89	30.60^0	0.97	57.07^0	1.39	51.05^0
ABC	2.07	44.97^0	1.42	26.36^0	1.41	50.04^0

TABLE 6.3

Classification Results from PNN

Type of Fault	No. of Events	PNN Output		% Correct
		No. of Correct Predictions	No. of Wrong predictions	
AG	84	84	0	100%
BG	84	84	0	100%
CG	84	84	0	100%
AB	84	84	0	100%
BC	84	81	3	96.4%
CA	84	84	0	100%
ABG	84	84	0	100%
BCG	84	84	0	100%
CAG	84	84	0	100%
ABC	84	84	0	100%

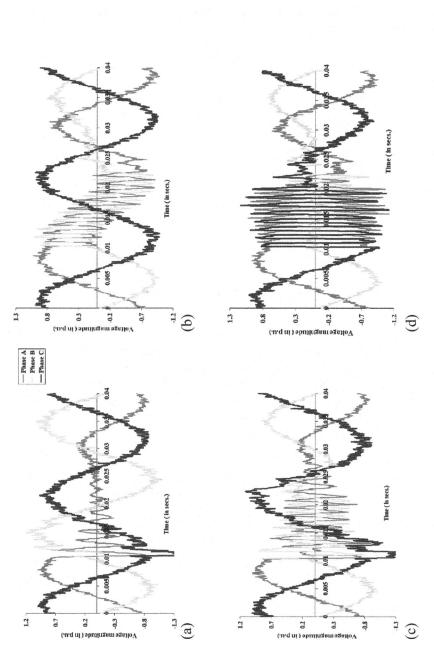

FIGURE 6.3 Voltage profiles of the three phases for (a) AG, (b) AB, (c) ABG, and (d) LLL type of faults at 100 km from B1, RF=0, θ = 00, and SNR=20 dB.

TABLE 6.4

Magnitudes of Classification Features for Different Types of Fault with Noise Added to the Current Signals

D=100 km, $R_F = 0$ Ω, $\theta = 0^0$	Q A	P A	Q B	P B	Q C	P C
PURE	1.01	57.83^0	1.00	58.39^0	1.00	57.48^0
AG	1.87	23.55^0	0.96	59.33^0	1.01	54.74^0
BG	1.03	55.63^0	1.93	24.58^0	0.97	57.94^0
CG	0.97	58.00^0	1.02	56.18^0	1.56	30.15^0
AB	1.69	67.65^0	1.59	16.66^0	1.01	57.36^0
BC	1.02	58.15^0	1.82	58.62^0	1.12	9.63^0
CA	1.14	35.50^0	1.00	58.40^0	1.41	56.36^0
ABG	1.90	27.35^0	2.23	31.72^0	2.23	55.62^0
BCG	0.97	56.37^0	1.93	28.65^0	1.54	26.67^0
CAG	1.93	22.76^0	0.96	56.83^0	1.61	31.44^0
ABC	1.87	49.94^0	2.21	35.05^0	1.62	29.83^0

6.7 FAULT LOCATION ESTIMATION BY BPNN

The absolute values of the S-matrices of the voltage signals measured at B1 for every phase have been analyzed. The study has revealed the presence of harmonics in the voltage signal of the faulty phase. The largest value occurring in each row of the S-matrix corresponds to the peak value of a particular harmonic. It has been seen from the simulations that a major frequency component remains constant with change in fault location but the next dominant frequency component changes with variation in fault location. The row numbers of the S-matrix corresponding to the variable frequency components have been used as input vectors for the training of the BPNN.

In this chapter, fault location has been determined by training the BPNN with Levenberg–Marquardt algorithm. The size of the input layer of the network in this paper is 15×10, where 15 represents the input features (i.e., fault locations) and ten represents the types of faults. The output of the network is the location of the particular fault, i.e., one real value. The back-propagation architecture has been tested with testing data set of 840 voltage signals. The results are given in Table 6.6, and they indicate that the fault location has been estimated with a maximum error of 4.46% without noise and 4.35% with noise added to the voltage signals. The errors in fault distance estimation are calculated using the following expression:

$$Error(\%) = \frac{BPNNoutput - T\arg et}{T\arg et} \times 100$$

TABLE 6.5
Results of Fault Location Estimation from BPNN

Fault Category	Fault Resistance (Ω)	Fault Location (km) Target Value	BPNN Output	% Error	Fault Category	Fault Resistance (Ω)	Fault Location (km) Target Value	BPNN Output	% Error
AG	0	20	20.65	3.25	CA	0	20	19.6	-2.00
		60	59.98	-0.03			60	61.78	2.97
		100	101.23	1.23			100	98.35	-1.65
		160	160.4	0.25			160	159.68	-0.20
		200	199.9	-0.05			200	200.99	0.50
		260	267.82	3.01			260	250	-3.85
		280	279.98	-0.01			280	273.47	-2.33
BG	0	20	19.96	-0.20	ABG	0	20	19.77	-1.15
		60	58.04	-3.27			60	61.28	2.13
		100	100.17	0.17			100	101.02	1.02
		160	162.8	1.75			160	158.24	-1.10
		200	205.49	2.75			200	207.23	3.61
		260	253.7	-2.42			260	265.29	2.03
		280	289.79	3.50			280	289.47	3.38
CG	0	20	20.69	3.45	BCG	0	20	20.72	3.60
		60	59.7	-0.50			60	58.92	-1.80
		100	100.04	0.04			100	101.39	1.39
		160	159.9	-0.06			160	159.7	-0.19
		200	199.69	-0.16			200	209.09	4.55
		260	254.13	-2.26			260	268.35	3.21
		280	279.8	-0.07			280	289.24	3.30

(Continued)

TABLE 6.5 (Continued)
Results of Fault Location Estimation from BPNN

Fault Category	Fault Resistance (Ω)	Fault Location (km)		% Error	Fault Category	Fault Resistance (Ω)	Fault Location (km)		% Error
		Target Value	BPNN Output				Target Value	BPNN Output	
AB	0	20	20.62	3.10	CAG	0	20	20.2	1.00
		60	60.38	0.63			60	58.06	-3.23
		100	99.2	-0.80			100	99.2	-0.80
		160	160.01	0.01			160	160.13	0.08
		200	199.76	-0.12			200	203.32	1.66
		260	252.55	-2.87			260	255.7	-1.65
		280	271.12	-3.17			280	277.79	-0.79
BC	0	20	20.53	2.65	ABC	0	20	20.95	4.75
		60	58.72	-2.13			60	61.27	2.12
		100	100.09	0.09			100	99.52	-0.48
		160	159.5	-0.31			160	160.23	0.14
		200	200.02	0.01			200	198.96	-0.52
		260	253.88	-2.35			260	265.37	2.07
		280	277.73	-0.81			280	271.16	-3.16
AG	20	20	20.70	3.50	CA	20	20	20.64	3.20
		60	60.00	0.00			60	62.06	3.43
		100	99.62	-0.38			100	101.27	1.27
		160	166.23	3.89			160	159.52	-0.30
		200	207.99	3.99			200	199.9	-0.05
		260	267.13	2.74			260	249.91	-3.88
		280	280.88	0.31			280	270.98	-3.22

Fault		Distance	Magnitude	Error
BG	20	20	19.85	-0.75
		60	57.94	-3.43
		100	102.04	2.04
		160	165.63	3.52
		200	204.8	2.40
		260	263.63	1.40
		280	289.89	3.53
CG	20	20	20.41	2.05
		60	60.72	1.20
		100	101.41	1.41
		160	163.95	2.47
		200	207.32	3.66
		260	265.54	2.13
		280	275.78	-1.51
AB	20	20	20.61	3.05
		60	57.9	-3.50
		100	98.86	-1.14
		160	159.78	-0.14
		200	200.83	0.42
		260	252.09	-3.04
		280	271.17	-3.15
BC	20	20	20.31	1.55
		60	59.53	-0.78
		100	99.7	-0.30
		160	157.85	-1.34
		200	200.48	0.24
		260	253.83	-2.37
		280	279.99	0.00

Fault		Distance	Magnitude	Error
ABG	20	20	19.27	-3.65
		60	62.23	3.72
		100	99.4	-0.60
		160	160.08	0.05
		200	203.32	1.66
		260	265.66	2.18
		280	288.58	3.06
BCG	20	20	20.59	2.95
		60	61.93	3.22
		100	101.36	1.36
		160	164.5	2.81
		200	208.91	4.46
		260	269.35	3.60
		280	289.91	3.54
CAG	20	20	19.99	-0.05
		60	58.17	-3.05
		100	99.58	-0.42
		160	161.14	0.71
		200	204.02	2.01
		260	255.99	-1.54
		280	279.99	0.00
ABC	20	20	20.55	2.75
		60	59.3	-1.17
		100	101.16	1.16
		160	160.08	0.05
		200	200.77	0.39
		260	250	-3.85
		280	277.85	-0.77

TABLE 6.6

Results of Fault Location Estimation from BPNN with 20 dB Noise Added to the Simulated Signals

Fault Category	Fault Resistance (Ω)	Fault Location (km)		% Error	Fault Category	Fault Resistance (Ω)	Fault Location (km)		% Error
		Target Value	BPNN Output				Target Value	BPNN Output	
AG	0	20	19.62	−1.90	CA	0	20	20.66	3.30
		60	60.21	0.35			60	61.61	2.68
		100	99.8	−0.20			100	99.24	−0.76
		160	160.14	0.09			160	160.08	0.05
		200	196.92	−1.54			200	199.98	−0.01
		260	269.7	3.73			260	249.91	−3.88
		280	280	0.00			280	278.3	−0.61
BG	0	20	20.46	2.30	ABG	0	20	19.54	−2.30
		60	61.52	2.53			60	59.18	−1.37
		100	98.37	−1.63			100	99.13	−0.87
		160	157.12	−1.80			160	159.18	−0.51
		200	199.13	−0.44			200	206.13	3.07
		260	249.99	−3.85			260	265.42	2.08
		280	287.49	2.68			280	288.67	3.10
CG	0	20	20.18	0.90	BCG	0	20	20.4	2.00
		60	62.34	3.90			60	61.88	3.13
		100	99.7	−0.30			100	99.45	−0.55
		160	160.18	0.11			160	159.45	−0.34
		200	199.72	−0.14			200	192.67	−3.67
		260	269.86	3.79			260	268.3	3.19
		280	280	0.00			280	289.2	3.29

Fault		Distance		
AB	0	20	20.79	3.95
		60	57.46	-4.23
		100	99.62	-0.38
		160	160.74	0.46
		200	200.82	0.41
		260	251.34	-3.33
		280	271.15	-3.16
BC	0	20	20.87	4.35
		60	59.11	-1.48
		100	99.52	-0.48
		160	159.08	-0.57
		200	199.31	-0.34
		260	253.53	-2.49
		280	279.98	-0.01
CAG	0	20	19.96	-0.20
		60	61.2	2.00
		100	99.1	-0.90
		160	159.21	-0.49
		200	203.88	1.94
		260	255.68	-1.66
		280	279.99	0.00
ABC	0	20	20.83	4.15
		60	64.79	7.98
		100	99.24	-0.76
		160	160.1	0.06
		200	199.89	-0.06
		260	250	-3.85
		280	272.69	-2.61

6.8 CONCLUSION

The feature extraction from the current or voltage signals is the most important part in fault detection using signal analysis. In this chapter, the number of features used for fault detection is six and for estimation of fault location the number is one. In the present system, the memory requirement and computation time is low. The proposed PNN classifier is quite robust and the results obtained are fast and accurate. The features can be conveniently obtained from the S-matrix by programming in MATLAB. The type of fault and the affected phase can be accurately identified using the proposed classification scheme. The PNN-based classifier has been rigorously tested by simulating fault conditions with different values of fault location, fault resistance, and fault inception angle. It has been found that the fault classifier can accurately classify the different types of faults with an average accuracy of 99.6%. The effect of noise on the simulated current signals has been also studied and the corresponding mean accuracy is 98.7%. On the other hand, the BPNN architecture proposed in this chapter can compute the fault location with a maximum error of 4.46%. The same BPNN is further investigated with simulated noisy voltage signals and the maximum error obtained is 4.35%. The results obtained indicate that the proposed method is capable of giving results with reasonably acceptable accuracy and speed.

6.9 FUTURE WORK FOR IMPLEMENTATION

The proposed technique would be tested in a multiterminal and multifeed network. The effect of transformers, CT, and PT would be studied by incorporating them in the present system. The results can be further investigated by using generalized ST and other fast versions of discrete ST.

6.A APPENDIX

Generator: Impedance of generator = (0.2+j4.49) Ω, X/R ratio = 22.45.
Transmission Line:

R_1 = 0.02336 Ω/km, R_2 = 0.02336 Ω/km, R_0 = 0.38848 Ω/km, L_1 = 0.95106 mH/km,

L_2 = 0.95106 mH/km, L_0 = 3.25083 mH/km, C_1 = 12.37 nF/km, C_2 = 12.37 nF/km,

C_0 = 8.45 nF/km.

Balanced Load: Load Impedance = (720+j11)Ω, P.f.= 0.9, MVA rating = 200

REFERENCES

1. M. V. Chilukuri and P. K. Dash, "Multiresolution S-Transform-Based Fuzzy Recognition System for Power Quality Events", *IEEE Transactions on Power Delivery*, vol. 19, no. 1, pp. 323–330, Jan 2004.

2. F. Zhao and R. Yang, "Power-quality disturbance recognition using S-transform", *IEEE Transactions on Power Delivery*, vol. 22, no. 2, pp. 944–950, Apr 2007.
3. S. Mishra, C. N. Bhende, and B. K. Panigrahi, "Detection and classification of power quality disturbances using S-transform and probabilistic neural network", *IEEE Transactions on Power Delivery*, vol. 23, no. 1, pp. 280–287, Jan 2008.
4. R. G. Stockwell, L. Mansinha, and R. P. Lowe, "Localization of the complex spectrum: The S-transform", *IEEE Transactions Signal Processing*, vol. 144, pp. 998–1001, Apr 1996.
5. A. Ngaopitakkul, and S. Bunjongjit, "An application of a discrete wavelet transform and a backpropagation neural network algorithm for fault diagnosis on single-circuit transmission line", *International Journal of Systems Science*, 2012. doi:10.1080/00207721.2012.670290.
6. Aritra Dasgupta, Sudipta Nath, and Arabinda Das, "Transmission line fault classification and location using wavelet entropy and neural network", *Electric Power Components and Systems*, vol. 40, no. 15, pp. 1676–1689, 2012.
7. A. Jamehbozorg and S. M. Shahrtash, "A decision-tree-based method for fault classification in single-circuit tramsmissiion lines", *IEEE Transactions on Power Delivery*, vol. 25, no. 4, pp. 2190–2195, Oct 2010.

7 Fault Analysis in an Unbalanced and a Multiterminal System Using ST and Neural Network

7.1 INTRODUCTION

Chapter 7 has demonstrated a technique for diagnosis of the type of fault and the faulty phase on overhead transmission line, followed by locating the particular fault on the affected phase. Various methods of Fault analysis based on Wavelet Transform, S-Transform and Neural Network have been provided in the literatures [1–21] and [25,26] and they have been thoroughly discussed in the Introductory and previous chapters. In this chapter, a similar technique has been implemented on an unbalanced power system network. The network considered in this study is a three-phase transmission line with unbalanced loading simulated in the PowerSim Toolbox of MATLAB. S-transform is used to compute energy components of the voltage signals of the three phases of the transmission line. These features are used as input vectors of a probabilistic Neural Network (PNN) for fault detection and classification. Detection of faulty phase(s) is followed by estimation of fault location. The voltage signal of the affected phase is processed to generate the S-matrix. The frequency components of the S-matrices for different fault locations are used as input vectors for training a Back Propagation Neural Network (BPNN). The results are obtained with satisfactory accuracy and speed. All the simulations have been done in MATLAB environment for different values of fault locations, fault resistances, and fault inception angles. The effect of noise on the simulated voltage signals has been investigated. The analysis has been further extended by implementing the proposed method in a modified version of IEEJ West 10 machine system model.

A PNN classifier based on ST has been proposed, where only three features (i.e., one feature/phase) are required for detecting a type of fault and the affected phase. The voltage signals of the three phases are processed through ST to generate complex S-matrices. The energy values of voltage signals of the three phases A, B, and C (EA, EB, and EC) are calculated from the absolute value of S-matrices. This simple feature extraction is done by programming in MATLAB. Since only three features are required, the memory requirement and computation time will significantly reduce. Moreover, using ST instead of WT will avoid the requirement of testing various families of wavelets in order to identify the best one for detection. The features extracted from ST are given to PNN for training, and subsequently, it is tested for an effective classification.

The rest of the chapter is organized as follows. The features of classification have been described in Section 7.2. Section 7.3 demonstrates a technique of fault classification and location estimation in an unbalanced power system network. The validity of the proposed technique has been further tested in a practical system in Section 7.4. A discussion has been included in Section 7.4 describing the difference of the proposed method with the other techniques based on ST and/or Artificial Neural network (ANN). The conclusion of this chapter followed by its future extension is presented in Section 7.5.

7.2   FEATURE EXTRACTION BY S-TRANSFORM

In the present analysis, ST has been employed to extract energy values of the voltage signals measured at sending end (B1) transmission line. The ST of the voltage signal of each phase would generate a complex matrix. Signal energy is obtained from the absolute value of the S-matrix.

Signal energy is calculated based on Parseval's Theorem. This theorem states that the energy of a signal remains the same whether it is computed in a signal domain (time) or in a transform domain (frequency) as given in Equation (7.1).

$$E_{signal} = \frac{1}{T}\int_{0}^{T}|v(t)|^2\,dt = \sum_{n=0}^{N}|V[n]|^2 \qquad (7.1)$$

where T and N are the time period and the length of the signal, respectively, and $V[n]$ is the Fourier transform of the signal.

In the case of the ST, the raw signal is decomposed in terms of its frequencies, and thus, a set of decomposed signals at each of the instantaneous frequencies in the raw signal can be obtained from the ST matrix. Thus, based on Parseval's Theorem, the energy of a distorted signal can be given as

$$E_{ST}\left[\frac{n}{NT}\right] = \sum_{k=1}^{N}\left(S\left[kT, \frac{n}{NT}\right]\right)^2 \qquad (7.2)$$

where $n = 1 \ldots \ldots \ldots N/2$, N is the signal length, and $E_{ST}\left[\dfrac{n}{NT}\right]$ is the energy vector of the instantaneous frequency at frequency n/NT.

In this article, the energy component of a voltage signal is calculated using Equation (7.3)

$$E = \sum_{i=1}^{M}Y(i)^2 \qquad (7.3)$$

where $Y(i)$=peak amplitude of STA matrix at i^{th} row of STA matrix and M = No. of sampled frequencies in the S-matrix. Each row of the STA matrix represents the instantaneous frequency.

Instead of a vector of data given by Equations (7.2, 7.3) gives a single numeric value and it efficiently serves the purpose of fault classification in the present work. This reduces the amount of input features needed for training a PNN and hence the time of classification.

7.3 FAULT CLASSIFICATION IN AN UNBALANCED POWER SYSTEM NETWORK

7.3.1 SIMULATION OF UNBALANCED SYSTEM

An unbalanced power system network is simulated using the Simpower Toolbox of MATLAB-7 and is shown in Figure 7.1. A three-phase unbalanced load is connected at the receiving end (B2) of a 400 kV, 50 Hz, three-phase transmission line. The length of the transmission line is 300 km. The parameters of the unbalanced load are given below:

Load 1: 180 MW, 87.2 MVAR

Load 2: 100 W, 70 MVAR

Load 3: 170 MW, 90 MVAR

The parameters of the transmission line have been given in the appendix. All the signals have been simulated with a sampling time of 78.28 μs, and the time period of simulation in MATLAB has been taken up to 0.04 secs. The sampling frequency is 12.8 kHz. All the ten types of faults have been simulated in this system as described in Chapter 5.

All the faults have been initiated at 29 different locations starting from B1, each being 10 km apart. The fault resistances considered for the simulation are from the range of 0 to 100 Ω in steps of 20 Ω. The faults have been initiated after 10 ms (i.e., one half cycle) and cleared at 20 ms. The fault inception angles are 0^0, 45^0, and 90^0. The total number of fault simulations made in this system is 10×29×6×3 = 5220.

7.3.2 PNN BASED FAULT CLASSIFICATION

The architecture of the PNN used for fault classification is same as described in Section 7.4 of Chapter 7 with the size of the input layer being 3×11, where 3

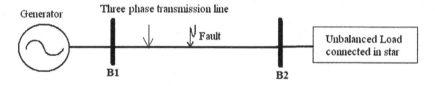

FIGURE 7.1   Single Line diagram of Three-Phase Network.

represents the input features (i.e., EA, EB, and EC of the three phases) and 11 is the number of fault types, including the no-fault condition. The normal signal and different faulty signals have been categorized as AG (1), BG (2), CG (3), AB (4), BC (5), CA (6), ABG (7), BCG (8), CAG (9), LLL (10), and normal (11), respectively. The size of the output layer of the PNN is 1×11. Out of 5220 cases of simulations, the features of ten faulty voltage signals for each type of fault have been used for training and the rest have been used for testing purposes.

7.3.3 Results of Simulation and PNN Classifier

Figure 7.2 represents the voltage waveforms for AG, AB, ABG, and LLL types of faults. Fast high-frequency oscillations are observed in the voltage profiles of the short-circuited phases during a fault condition. The amplitude of the shorted phases at the instant of fault occurrence, however, depends on the time of initiation of the fault. In this chapter, the time of occurrence of the fault as mentioned before is taken at three different instants, i.e., after 10 ms, (i.e., after one half cycle), 12.5 ms, and 15 ms. The amplitudes of sending end phase voltages of A, B, and C after 10 ms are 129.26 kV, 48.80 kV, and −178.06 kV, respectively.

Table 7.1 shows the values of classification features of all the fault types that have occurred at distances of 10 km and 180 km from the sending end of the transmission line (B1). The results of the PNN classifier are shown in Table 7.2. It is noticed in Table 7.2, that the total no. of misclassified cases is 19. The average accuracy of classification in the present study is 99.6%.

7.3.4 Effect of Noise

White gaussian noise has been added to the simulated voltage signals by considering a noise level of 20 dB SNR. The noisy voltage profiles in the case of AG, AB, ABG, and LLL faults are shown in Figure 7.3.

The entire testing data has been impregnated with this noise by simulation in MATLAB. As an illustration, Table 7.3 shows the values of signal energy for a particular fault condition in noisy environment. The average accuracy of fault classification obtained is 96.8%.

7.3.5 Fault Location Estimation by BPNN

The Levenberg–Marquardt algorithm has been used to train the BPNN for obtaining the fault location. The size of the input layer of the network is 15×10, where 15 represents the input features (i.e., fault locations) and 10 represents the types of faults. The BPNN has been tested with 512 voltage signals. The results are summarized in Tables 7.4–7.6, and they indicate that the fault location has been estimated with a maximum error of 4.75% without noise and 4.27% with noise added to the voltage signals. The errors in fault distance estimation have been calculated using the following expression:

$$Error\,(\%) = \frac{BPNNoutput - T\arg et}{T\arg et} \times 100$$

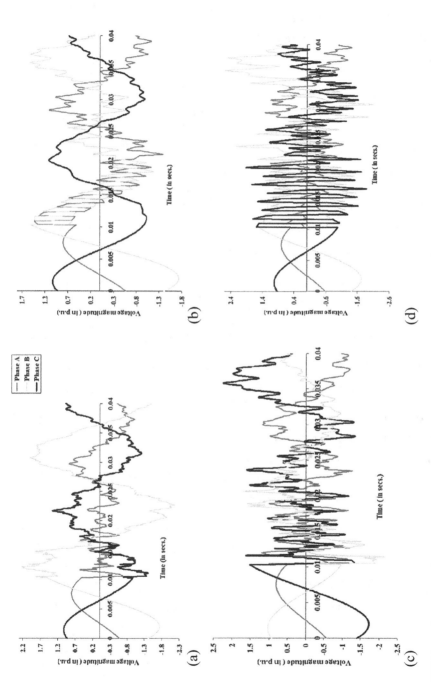

FIGURE 7.2 Voltage profiles of the three phases for (a) AG, (b) AB, (c) ABG, and (d) LLL type of faults at 100 km from B1, $R_F = 0$, and $\theta = 0^0$.

TABLE 7.1
Magnitudes of Signal Energy for Different Types of Fault

D=10km, RF = 0 Ω, θ = 0⁰	EA (p.u.)	EB (p.u.)	EC (p.u.)
NORMAL	0.16	0.76	0.27
AG	1.32	1.09	0.50
BG	1.50	10.98	1.62
CG	0.83	1.33	3.78
AB	0.80	1.19	0.26
BC	0.22	10.68	10.15
CA	2.97	0.75	3.07
ABG	4.27	6.91	1.78
BCG	0.28	11.56	7.52
CAG	3.14	0.80	4.89
LLL	0.55	8.26	11.14
D=290km, RF = 0 Ω, θ = 0⁰	EA (p.u.)	EB (p.u.)	EC(p.u.)
NORMAL	0.16	0.76	0.27
AG	0.60	1.02	0.47
BG	1.15	2.30	1.24
CG	0.53	1.07	0.92
AB	0.52	0.93	0.27
BC	0.27	3.48	2.97
CA	1.22	0.75	1.30
ABG	1.47	2.09	2.48
BCG	0.43	3.61	2.85
CAG	1.20	0.77	1.30
LLL	0.25	2.81	3.40

TABLE 7.2
Classification Results from PNN

Type of Fault	No. of Events	PNN Output		% Correct
		No. of Correct Predictions	No. of Wrong Predictions	
AG	512	500	12	97.6%
BG	512	512	0	100%
CG	512	512	0	100%
AB	512	510	2	99.6%
BC	512	512	0	100%
CA	512	512	0	100%
ABG	512	510	2	99.6%
BCG	512	509	3	99.4%
CAG	512	512	0	100%
ABC	512	512	0	100%

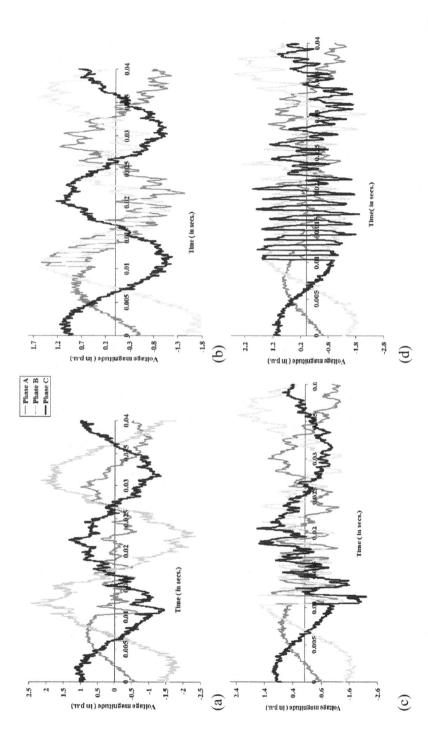

FIGURE 7.3 Voltage profiles of the three phases for (a) AG, (b) AB, (c) ABG, and (d) LLL type of faults at 100 km from B1, RF = 0, θ = 00, and SNR=20 dB.

TABLE 7.3
Magnitudes of Classification Features with Noise Added to the Voltage Signals

D=290 km, RF = 0 Ω, θ = 0⁰	EA (p.u.)	EB (p.u.)	EC (p.u.)
AG	0.72	1.61	0.68
BG	1.21	2.36	1.21
CG	0.60	1.54	1.09
AB	0.66	1.25	0.46
BC	0.38	4.17	3.27
CA	1.35	1.51	1.60
ABG	1.67	2.38	2.14
BCG	0.47	3.47	3.35
CAG	1.19	1.46	1.38
LLL	0.37	3.09	3.62

7.4   APPLICATION OF THE PROPOSED METHOD IN A PRACTICAL SYSTEM

A modified version of the IEEJ West 10 machine System model [22] has been simulated in MATLAB 7 environment. Each of the ten machines has been modeled as a three- phase voltage source available in the library of MATLAB software. The rating of the machine is given as provided in the Table of the IEEJ West model [23]. Each of the transmission line has been modeled as a single-circuit three-phase distributed line available in the MATLAB toolbox. The transmission line parameters are the same as those of the system in Figure 7.1. Hence, there are 16 transmission lines (branches) in the simulated network. Each of the interconnected line is of 100 km, and the length of each line connected to the generators is 50 km (excepting the one connected to G8), as per the data given in Table of the IEEJ West model [24] (Figure 7.4).

All the ten types of faults have been simulated on the 16 lines at different locations in steps of 10 km starting from B1, B2, B3, B4, B5, B6, B7, and B8, respectively. Seven lines are of 100 km and the remaining nine lines are of 50 km. Fault resistance is 0 Ω, and the fault inception angle is 0⁰. Hence, the total number of simulations is 10×9×7+10×4×9 = 990. The voltage waveforms for particular fault conditions are shown in Figure 7.5.

The features of 15 voltage signals for each type of fault have been used for training the PNN. The Table 7.7 shows the energy values for all the 16 lines at a particular location for AG and BG types of fault.

The results of classification are tabulated in Table 7.8. It is observed that there are 24 wrong predictions in Table 7.8 and the average accuracy of classification achieved is 97.6%.

TABLE 7.4
Results of Fault Location Estimation from BPNN with Fault Resistance 0 Ω

Fault Category	Fault Resistance (Ω)	Fault Location (km) Target Value	BPNN Output	% Error	Fault Category	Fault Resistance (Ω)	Fault Location (km) Target Value	BPNN Output	% Error
AG	0	20	20.65	3.25	CA	0	20	19.6	-2.00
		60	59.98	-0.03			60	61.78	2.97
		100	101.23	1.23			100	98.35	-1.65
		160	160.4	0.25			160	159.68	-0.20
		200	199.9	-0.05			200	200.99	0.50
		260	267.82	3.01			260	250	-3.85
		280	279.98	-0.01			280	273.47	-2.33
BG	0	20	19.96	-0.20	ABG	0	20	19.77	-1.15
		60	58.04	-3.27			60	61.28	2.13
		100	100.17	0.17			100	101.02	1.02
		160	162.8	1.75			160	158.24	-1.10
		200	205.49	2.75			200	207.23	3.61
		260	253.7	-2.42			260	265.29	2.03
		280	289.79	3.50			280	289.47	3.38
CG	0	20	20.69	3.45	BCG	0	20	20.72	3.60
		60	59.7	-0.50			60	58.92	-1.80
		100	100.04	0.04			100	101.39	1.39
		160	159.9	-0.06			160	159.7	-0.19
		200	199.69	-0.16			200	209.09	4.55
		260	254.13	-2.26			260	268.35	3.21
		280	279.8	-0.07			280	289.24	3.30
AB	0	20	20.62	3.10	CAG	0	20	20.2	1.00
		60	60.38	0.63			60	58.06	-3.23
		100	99.2	-0.80			100	99.2	-0.80
		160	160.01	0.01			160	160.13	0.08

(Continued)

TABLE 7.4 (Continued)
Results of Fault Location Estimation from BPNN with Fault Resistance 0 Ω

Fault Category	Fault Resistance (Ω)	Fault Location (km) Target Value	BPNN Output	% Error	Fault Category	Fault Resistance (Ω)	Fault Location (km) Target Value	BPNN Output	% Error
BC	0	200	199.76	−0.12	ABC	0	200	203.32	1.66
		260	252.55	−2.87			260	255.7	−1.65
		280	271.12	−3.17			280	277.79	−0.79
		20	20.53	2.65			20	20.95	4.75
		60	58.72	−2.13			60	61.27	2.12
		100	100.09	0.09			100	99.52	−0.48
		160	159.5	−0.31			160	160.23	0.14
		200	200.02	0.01			200	198.96	−0.52
		260	253.88	−2.35			260	265.37	2.07
		280	277.73	−0.81			280	271.16	−3.16

TABLE 7.5

Results of Fault Location Estimation from BPNN with Fault Resistance 20 Ω

Fault Category	Fault Resistance (Ω)	Fault Location (km) Target Value	BPNN Output	% Error	Fault Category	Fault Resistance (Ω)	Fault Location (km) Target Value	BPNN Output	% Error
AG	20	20	20.70	3.50	CA	20	20	20.64	3.20
		60	60.00	0.00			60	62.06	3.43
		100	99.62	-0.38			100	101.27	1.27
		160	166.23	3.89			160	159.52	-0.30
		200	207.99	3.99			200	199.9	-0.05
		260	267.13	2.74			260	249.91	-3.88
		280	280.88	0.31			280	270.98	-3.22
BG	20	20	19.85	-0.75	ABG	20	20	19.27	-3.65
		60	57.94	-3.43			60	62.23	3.72
		100	102.04	2.04			100	99.4	-0.60
		160	165.63	3.52			160	160.08	0.05
		200	204.8	2.40			200	203.32	1.66
		260	263.63	1.40			260	265.66	2.18
		280	289.89	3.53			280	288.58	3.06
CG	20	20	20.41	2.05	BCG	20	20	20.59	2.95
		60	60.72	1.20			60	61.93	3.22
		100	101.41	1.41			100	101.36	1.36
		160	163.95	2.47			160	164.5	2.81
		200	207.32	3.66			200	208.91	4.46
		260	265.54	2.13			260	269.35	3.60
		280	275.78	-1.51			280	289.91	3.54

(Continued)

TABLE 7.5 (Continued)

Results of Fault Location Estimation from BPNN with Fault Resistance 20 Ω

Fault Category	Fault Resistance (Ω)	Fault Location (km) Target Value	BPNN Output	% Error	Fault Category	Fault Resistance (Ω)	Fault Location (km) Target Value	BPNN Output	% Error
AB	20	20	20.61	3.05	CAG	20	20	19.99	-0.05
		60	57.9	-3.50			60	58.17	-3.05
		100	98.86	-1.14			100	99.58	-0.42
		160	159.78	-0.14			160	161.14	0.71
		200	200.83	0.42			200	204.02	2.01
		260	252.09	-3.04			260	255.99	-1.54
		280	271.17	-3.15			280	279.99	0.00
BC	20	20	20.31	1.55	ABC	20	20	20.55	2.75
		60	59.53	-0.78			60	59.3	-1.17
		100	99.7	-0.30			100	101.16	1.16
		160	157.85	-1.34			160	160.08	0.05
		200	200.48	0.24			200	200.77	0.39
		260	253.83	-2.37			260	250	-3.85
		280	279.99	0.00			280	277.85	-0.77

TABLE 7.6

Results of Fault Location Estimation from BPNN with 20 dB Noise Added to the Voltage Signals

Fault Category	Fault Resistance (Ω)	Fault Location (km)		% Error	Fault Category	Fault Resistance (Ω)	Fault Location (km)		% Error
		Target Value	BPNN Output				Target Value	BPNN Output	
AG	0	20	19.62	−1.90	CA	0	20	20.66	3.30
		60	60.21	0.35			60	61.61	2.68
		100	99.8	−0.20			100	99.24	−0.76
		160	160.14	0.09			160	160.08	0.05
		200	196.92	−1.54			200	199.98	−0.01
		260	269.7	3.73			260	249.91	−3.88
		280	280	0.00			280	278.3	−0.61
BG	0	20	20.46	2.30	ABG	0	20	19.54	−2.30
		60	61.52	2.53			60	59.18	−1.37
		100	98.37	−1.63			100	99.13	−0.87
		160	157.12	−1.80			160	159.18	−0.51
		200	199.13	−0.44			200	206.13	3.07
		260	249.99	−3.85			260	265.42	2.08
		280	287.49	2.68			280	288.67	3.10
CG	0	20	20.18	0.90	BCG	0	20	20.4	2.00
		60	62.34	3.90			60	61.88	3.13
		100	99.7	−0.30			100	99.45	−0.55
		160	160.18	0.11			160	159.45	−0.34
		200	199.72	−0.14			200	192.67	−3.67
		260	269.86	3.79			260	268.3	3.19
		280	280	0.00			280	289.2	3.29

(Continued)

TABLE 7.6 (Continued)

Results of Fault Location Estimation from BPNN with 20 dB Noise Added to the Voltage Signals

Fault Category	Fault Resistance (Ω)	Fault Location (km) Target Value	BPNN Output	% Error	Fault Category	Fault Resistance (Ω)	Fault Location (km) Target Value	BPNN Output	% Error
AB	0	20	20.79	3.95	CAG	0	20	19.96	-0.20
		60	57.46	-4.23			60	61.2	2.00
		100	99.62	-0.38			100	99.1	-0.90
		160	160.74	0.46			160	159.21	-0.49
		200	200.82	0.41			200	203.88	1.94
		260	251.34	-3.33			260	255.68	-1.66
		280	271.15	-3.16			280	279.99	0.00
BC	0	20	20.87	4.35	ABC	0	20	19.89	-0.55
		60	59.11	-1.48			60	62.56	4.27
		100	99.52	-0.48			100	99.24	-0.76
		160	159.08	-0.57			160	160.1	0.06
		200	199.31	-0.34			200	199.89	-0.06
		260	253.53	-2.49			260	250	-3.85
		280	279.98	-0.01			280	272.69	-2.61

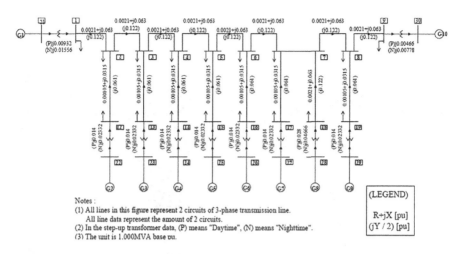

FIGURE 7.4 IEEJ West 10 Model.

Fault location estimation has been achieved in two steps:

- Identification of the particular transmission line segment in which the fault has occurred.
- Determination of the exact location of the fault in the identified faulty section.

7.4.1 Identification of Faulty Line Segment

A new PNN has been developed to identify the faulty line section. The PNN architecture remains the same as described in Section 7.3.2. Only the size of the input and output layers has changed. The energy values of four voltage signals of only the faulty phase corresponding to each interconnecting line between two buses of the IEEJ West 10 machine system model have been used as the input vector of the PNN. The size of the input layer for the PNN is 16×10, where 16 represents the input features (i.e., energy value of the faulty phase of every interconnecting line) and 10 is the number of fault types. The transmission lines between the buses have been categorized as Line1–2 (1), Line 2–3 (2), Line 3–4 (3), Line 4–5 (4), Line 5–6 (5), Line 6–7 (6), Line 7–8 (7), Line 8–9 (8), Line2–12 (9), Line3–13 (10), Line 4–14 (11), Line 5–15 (12), Line 6–16 (13), Line 7–17 (14), Line 7–18 (15), and Line 8–19 (16), respectively. The size of the output layer of the PNN is 1×16, and the output is an integer ranging from one to 16 indicating the line section between two buses in which the fault has occurred. Fault signals have been simulated in steps of 15 km in every interconnected line starting from B1, B2, B3, B4, B5, B6, B7, and B8, respectively. The values of fault resistance and fault inception angle are 0 Ω and 0°, respectively. The total number of simulations have

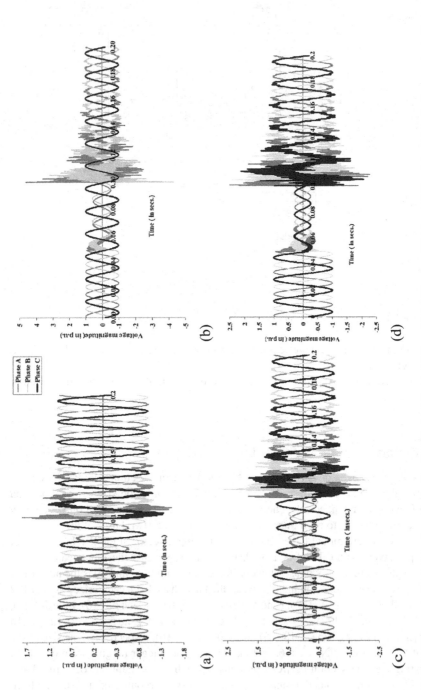

FIGURE 7.5　Voltage profiles of the three phases for (a) AG, (b) AB, (c) ABG, and (d) LLL type of faults at 20 km from B1 in the line connected between B1 and B2 with $R_F = 0$ and $\theta = 0^\circ$.

TABLE 7.7

Magnitudes of Signal Energy for AG and BG Types of Fault on All the Interconnecting Lines between the Buses

Type of Fault	Fault Location	EA (p.u.)	EB (p.u.)	EC (p.u.)
No Fault	On every line	0.43	0.43	0.43
AG	Line 1–2: 10 km from B1	15.05	4.41	4.42
	Line 2–3: 10 km from B2	2.45	0.89	0.90
	Line 3–4: 10 km from B3	0.79	0.52	0.53
	Line 4–5: 10 km from B4	0.76	0.53	0.53
	Line 5–6: 10 km from B5	0.96	0.57	0.57
	Line 6–7: 10 km from B6	1.30	0.65	0.65
	Line 7–8: 10 km from B7	0.81	0.53	0.53
	Line 8–9: 10 km from B8	0.44	0.43	0.43
	Line2–12: 10 km from B2	3.32	1.16	1.16
	Line3–13: 10 km from B3	0.73	0.51	0.51
	Line4–14: 10 km from B4	0.76	0.53	0.53
	Line5–15: 10 km from B5	1.03	0.59	0.59
	Line6–16: 10 km from B6	1.42	0.69	0.69
	Line7–17: 10 km from B7	0.61	0.49	0.48
	Line7–18: 10 km from B7	1.10	0.61	0.61
	Line8–19: 10 km from B8	0.49	0.45	0.45
BG	Line 1–2: 10 km from B1	15.01	47.72	15.00
	Line 2–3: 10 km from B2	7.44	29.11	7.44
	Line 3–4: 10 km from B3	0.97	2.50	0.97
	Line 4–5: 10 km from B4	0.83	1.78	0.82
	Line 5–6: 10 km from B5	0.70	1.44	0.70
	Line 6–7: 10 km from B6	0.63	1.25	0.63
	Line 7–8: 10 km from B7	0.47	0.56	0.47
	Line 8–9: 10 km from B8	0.44	0.47	0.44
	Line2–12: 10 km from B2	4.57	16.72	4.57
	Line3–13: 10 km from B3	0.77	1.84	0.77
	Line4–14: 10 km from B4	0.80	1.65	0.80
	Line5–15: 10 km from B5	0.79	1.79	0.80
	Line6–16: 10 km from B6	0.69	1.43	0.69
	Line7–17: 10 km from B7	0.48	0.62	0.49
	Line7–18: 10 km from B7	0.49	0.63	0.49
	Line8–19: 10 km from B8	0.44	0.46	0.44

TABLE 7.8

Classification Results from PNN for the IEEJ West 10 Machine System Model

Type of Fault	No. of Events	PNN Output		% Correct
		No. of Correct Predictions	No. of Wrong Predictions	
AG	99	98	1	98.9%
BG	99	97	2	97.9%
CG	99	98	1	98.9%
AB	99	95	4	95.9%
BC	99	97	2	97.9%
CA	99	98	1	98.9%
ABG	99	95	4	95.9%
BCG	99	96	3	96.9%
CAG	99	97	2	97.9%
ABC	99	95	4	95.9%

TABLE 7.9

Results of Faulty Line Segment Identification from PNN

Type of Fault	No. of Events	PNN Output		% Correct
		No. of Correct Identifications	No. of Wrong Identifications	
AG	69	68	1	98.6%
BG	69	65	4	94.2%
CG	69	68	1	98.6%
AB	69	66	3	95.6%
BC	69	65	4	94.2%
CA	69	66	3	95.6%
ABG	69	67	2	97.1%
BCG	69	68	1	98.6%
CAG	69	67	2	97.1%
ABC	69	69	0	100%

been $10 \times 6 \times 7 + 10 \times 3 \times 9 = 690$, and they have been used for testing the PNN. The results are tabulated in Table 7.9, and the number of wrong identifications is 21. The average accuracy of correct identification is 96.9%.

7.4.2 DETERMINATION OF EXACT FAULT LOCATION

The BPNN used for computation of fault location has the same architecture as described in Section 6. The frequency components of nine voltage signals of the faulty phase out of 990 simulations have been taken as inputs for training this BPNN. The size of the input layer of the network is 9×10. This BPNN has been tested with the voltage signals against which the faulty line segments have been correctly identified in Table 7.9. The results have been summarized in Table 7.10, and they indicate that the fault location has been estimated with a maximum error of 4.13%.

TABLE 7.10

Results of Fault Location Estimation from BPNN for IEEJ West 10 Machine System Model

Fault Category	Fault Line Segment	Fault Location (km) Target Value	BPNN Output	% Error	Fault Category	Faulty Line Segment	Fault Location (km) Target Value	BPNN output	% Error
AG	Line 1–2	15	15.02	0.13	CA	Line 1–2	15	15.36	2.40
	Line 2–3	60	58.67	-2.22		Line 2–3	60	59.97	-0.05
	Line 3–4	75	75.12	0.16		Line 3–4	75	74.89	-0.15
	Line 4–5	90	90.23	0.26		Line 4–5	90	90.12	0.13
	Line 2–12	30	28.67	-4.43		Line 2–12	30	30	0.00
	Line 3–13	45	44.32	-1.51		Line 3–13	45	45.11	0.24
	Line 4–14	15	14.87	-0.87		Line 4–14	15	15.02	0.13
BG	Line 1–2	15	15	0.00	ABG	Line 1–2	15	15	0.00
	Line 2–3	60	59.99	-0.02		Line 2–3	60	60.04	0.07
	Line 3–4	75	74.56	-0.59		Line 3–4	75	75	0.00
	Line 4–5	90	89.12	-0.98		Line 4–5	90	90.8	0.89
	Line 2–12	30	29	-3.33		Line 2–12	30	30.5	1.67
	Line 3–13	45	44.34	-1.47		Line 3–13	45	45.32	0.71
	Line 4–14	15	14.89	-0.73		Line 4–14	15	14.89	-0.73
CG	Line 1–2	15	15.62	4.13	BCG	Line 1–2	15	15.82	1.53
	Line 2–3	60	59.89	-0.18		Line 2–3	60	59.9	-0.17
	Line 3–4	75	75	0.00		Line 3–4	75	74.33	-0.89
	Line 4–5	90	89.34	-0.73		Line 4–5	90	88.98	-1.13
	Line 2–12	30	29.56	-1.47		Line 2–12	30	30	0.00
	Line 3–13	45	45.1	0.22		Line 3–13	45	44.7	-0.67
	Line 4–14	15	15.1	0.67		Line 4–14	15	15	0.00

(Continued)

TABLE 7.10 (Continued)
Results of Fault Location Estimation from BPNN for IEEJ West 10 Machine System Model

Fault Category	Fault Line Segment	Fault Location (km) Target Value	BPNN Output	% Error	Fault Category	Faulty Line Segment	Fault Location (km) Target Value	BPNN output	% Error
AB	Line 1–2	15	15.23	1.53	CAG	Line 1–2	15	15.2	1.33
	Line 2–3	60	59.23	−1.28		Line 2–3	60	59.12	−1.47
	Line 3–4	75	75	0.00		Line 3–4	75	75.64	0.85
	Line 4–5	90	89.23	−0.86		Line 4–5	90	89.9	−0.11
	Line 2–12	30	30	0.00		Line 2–12	30	30	0.00
	Line 3–13	45	45.02	0.04		Line 3–13	45	45.4	0.89
	Line 4–14	15	15	0.00		Line 4–14	15	15	0.00
BC	Line 1–2	15	14.89	−0.73	ABC	Line 1–2	15	14.89	−0.73
	Line 2–3	60	59.9	−0.17		Line 2–3	60	58.89	−1.85
	Line 3–4	75	75.45	0.60		Line 3–4	75	75	0.00
	Line 4–5	90	89.98	−0.02		Line 4–5	90	89.9	−0.11
	Line 2–12	30	30.15	0.50		Line 2–12	30	29.98	−0.07
	Line 3–13	45	45.25	0.56		Line 3–13	45	45.56	1.24
	Line 4–14	15	14.89	−0.73		Line 4–14	15	14.82	−1.20

7.5 CONCLUSION

In the present chapter, a combination of PNN and BPNN has been proposed for fault classification, faulty section identification (in case of multiterminal system), and determination of fault location. LM is a training algorithm suitable for BPNN network and has the ability of faster convergence compared to other back propagation training algorithms [27]. But BPNN is relatively more effective in solving function approximation problems where a desirable output is required [27]. PNN can be trained to be a relatively stronger classifier and its architecture is very simple [28, 29]. The advantage of PNN is its requirement of few training samples and its flexibility of adding new information in the input set without requiring any additional training [28, 29]. Both PNN and BPNN have their own limitations and advantages. Application of both these methods has produced acceptable results in the present study.

The proposed method has been applied on a 2-bus system and a multiterminal practical system consisting of 27 terminals and 16 transmission lines. The total number of fault simulations is 5220 for the first system. In the second system (990+690=) 1680 fault conditions have been simulated for training and testing the neural networks in different ways. The effect of noise on the simulated voltage signals has been studied in detail. The results have been obtained with acceptable accuracy.

The feature extraction from the current or voltage signals is the most important part in fault detection using signal analysis. The proposed method is capable of making fault identification and location estimation and speed. Only energy component of the voltage signals of the three phases have been considered for fault classification, and harmonic component of faulty voltage waveform has been selected for fault location estimation. Hence, the size of the feature vectors has reduced considerably. The PNN architecture-based fault category identification and the BPNN architecture-based fault distance estimation proposed in this article has reasonably acceptable accuracy by simulating fault conditions with different values of fault location, fault resistance, and fault inception angle. The effect of noise has been investigated by simulating the voltage signals with White Gaussian Noise. It has been found that the fault classifier can classify the different types of faults in unbalanced loading with 99.6% accuracy for pure signals and 96.8% accuracy for noisy signals. The fault distance estimation has been possible with a maximum error of 4.75% without noise and 4.27% for noisy signals.

The validity of the proposed method is further tested by implementing it in a modified IEEJ West 10 machine system model. The faults have been classified in the system with an average accuracy of 97.6%. As this is a multiterminal system, fault location estimation has been performed by a combination of PNN and BPNN architectures. The faults have been located with a maximum error of 4.13%.

The results obtained indicate that this method is capable of detecting all types of faults along with affected phases and estimating distance of fault location with acceptable accuracy.

For a larger system with enormous data, a fast version of discrete ST is recommended to be implemented in future. As the simulation of the actual IEEJ West 10 machine system model requires an advance version of MATLAB/EMTP software the accuracy of the present technique is intended to be further tested on the actual model where all the interconnecting lines are double-circuit systems.

REFERENCES

1. P. S. Bhowmik, P. Purkait, and K. Bhattacharya, "A novel wavelet transform aided neural network based transmission line fault analysis method", *International Journal of Electrical Power and Energy Systems*, vol. 31, no. 5, pp. 213–219, Jun 2009.
2. A. Ngaopitakkul, and S. Bunjongjit, "An application of a discrete wavelet transform and a backpropagation neural network algorithm for fault diagnosis on single-circuit transmission line", *International Journal of Systems Science*, 2012. doi: 10.1080/00207721.2012.670290.
3. Bhavesh Bhalja, and R. P. Maheshwari, "Wavelet-based fault classification scheme for a transmission line using a support vector machine", *Electric Power Components and Systems*, vol. 36, no. 10, pp. 1017–1030, 2008.
4. O.A.S. Youssef, "Combined fuzzy-logic wavelet –based fault classification technique for power system relaying", *IEEE Transactions on Power Delivery*, vol. 19, no. 2, pp. 582–589, Apr 2004.
5. Thai Nguyen, and Yuan Liao, "Transmission line fault type classification based on novel features and neuro-fuzzy system", *Electric Power Components and Systems*, vol. 38, no. 6, pp. 695–709, 2010.
6. Theerasak Patcharoen and Atthapol Ngaopitakkul, "A novel discrete wavelet transform based on traveling wave technique for identifying the fault location for transmission network systems", *IEEJ Transactions on Electrical and Electronic Engineering*, vol. 8, no. 5, Sep 2013, pp. 432–439, doi: 10.1002/tee.21877.
7. Tai NengLing and Chen JiaJia, "Wavelet-based approach for high mpedance fault detection of high voltage transmission line", *European Transactions on Electrical Power*, vol. 18, no. 1, pp. 79–92, Jan 2008.
8. Mohd Syukri Ali, Ab Halim Abu Bakar, Hazlie Mokhlis, Hamzah Arof, and Hazlee Azil Illias, "High-impedance fault location using matching technique and wavelet transform for underground cable distribution network", *IEEJ Transactions on Electrical and Electronic Engineering*, vol. 9, no. 2, pp. 176–182, Mar 2014. doi: 10.1002/tee.21953.
9. Aritra Dasgupta, Sudipta Nath, Arabinda Das, "Transmission line fault classification and location using wavelet entropy and neural network", *Electric Power Components and Systems*, vol. 40, no. 15, pp. 1676–1689, 2012.
10. M. V. Chilukuri and P. K. Dash, "Multiresolution S-transform-based fuzzy recognition system for power quality events", *IEEE Transaction on Power Delivery*, vol. 19, no. 1, pp. 323–330, Jan 2004.
11. S. Mishra, C. N. Bhende, and B. K. Panigrahi, "Detection and classification of power quality disturbances using S-transform and probabilistic neural network", *IEEE Transactions on Power Delivery*, vol. 23, no. 1, pp. 280–287, Jan 2008.
12. C. Venkatesh, D. V. S. S. Siva Sarma, and M. Sydulu, "Detection of power quality disturbances using phase corrected wavelet transform", *Journal of The Institution of Engineers (India): Series*, vol. 93, no. 1, pp. 37–42, Mar–May 2012. doi: 10.1007/s40031-012-0006-z.
13. N. Perera and A. D. Rajapakse, "Recognition of fault transients using a probabilistic neural-network classifier", *IEEE Transactions on Power Delivery*, vol. 26, no. 1, pp. 410–419, Jan 2011.

14. Maryam Mirzaei, Mohd Zainal Abidin Ab. Kadir, Hashim Hizam, and Ehsan Moazami, "Comparative analysis of probabilistic neural network, radial basis function, and feed-forward neural network for fault classification in power distribution systems", *Electric Power Components and Systems*, vol. 39, no. 16, pp. 1858–1871, 2011.

15. S. F. Mekhamer, A. Y. Abdelaziz, M. Ezzat, and T. S. Abdel-Salam, "Fault location in long transmission lines using synchronized phasor measurements from both ends", *Electric Power Components and Systems*, vol. 40, no. 7, pp. 759–776, 2012.

16. Yuan Liao, "Fault location utilizing unsynchronized voltage measurements during fault", *Electric Power Components and Systems*, vol. 34, no. 12, pp. 1283–1293.

17. Shamam Fadhil Alwash and Vigna Kumaran Ramachandaramurthy, "Novel fault-location method for overhead electrical distribution systems", *IEEJ Transactions on Electrical and Electronic Engineering*, vol. 8, no. S1, pp. S13–S19, 2013, doi: 10.1002/tee.21913.

18. Lilik Jamilatul Awalin, Hazlie Mokhlis, AbHalim Abu Bakar, Hasmaini Mohamad, and Hazlee A. Illias, "A generalized fault location method based on voltage sags for distribution network", *IEEJ Transactions on Electrical and Electronic Engineering*, vol. 8, no. S1, pp. S38–S46, 2013. doi: 10.1002/tee.21916.

19. Wutthikorn Threevithayanon and Naebboon Hoonchareon, "Accurate one-terminal fault location algorithm based on the principle of short-circuit calculation", *IEEJ Transactions on Electrical and Electronic Engineering*, vol. 8, no. 1, pp. 28–32, Jan 2013. doi: 10.1002/tee.21787.

20. Y. Wang and J. Orchard, "Fast discrete orthonormal stockwell transform", *SIAM Journal on Scientific Computing*, vol. 31, no. 5, pp. 4000–4012, 2009.

21. M. Biswal, and P.K. Dash, "Estimation of time-varying power quality indices with an adaptive window-based fast generalised S-transform", *IET Science, Measurement & Technology*, vol. 6, no. 4, pp. 189–197, 2012.

22. www2.iee.or.jp/~pes/model/english/kikan/West10/west10.html.

23. www2.iee.or.jp/~pes/model/english/kikan/West10/tab.3.1.pdf.

24. H. Shu, Q. Wu, X. Wang and X. Tian, "Fault phase selection and distance location based on ANN and S-transform for transmission line in triangle network," *2010 3rd International Congress on Image and Signal Processing*, Yantai, China, 2010, pp. 3217–3219, doi: 10.1109/CISP.2010.5648109.

25. K. R. Krishnanand, P. V. Balasubramanyam, S. K. Swain and P. K. Dash, "S-transform based spectral energy feature space for fault location approximation," *2011 International Conference on Energy, Automation and Signal*, Bhubaneswar, India, 2011, pp. 1–5, doi: 10.1109/ICEAS.2011.6147108.

26. Alireza Ahmadimanesh and S. Mohammad Shahrtash, "Transient-based fault-location method for multiterminal lines employing S-transform", *IEEE Transactions on Power Delivery*, vol. 28, no. 3, pp. 1373–1380, Jul 2013.

27. www.mathworks.in/help/pdf_doc/nnet/nnet_ug.pdf.

28. https://minds.wisconsin.edu/bitstream/handle/1793/7779/ch1_3.pdf.

29. Mustafa Ghaderzadeh, Rebecca Fein, and Arran Standring, "Comparing performance of different neural networks for early detection of cancer from benign hyperplasia of prostate", *Applied Medical Informatics*, vol. 33, no. 3, pp. 45–54, 2013.

8 Application of ST for Fault Analysis in a HVDC System

8.1 INTRODUCTION

With the development of power system, HVDC systems play important roles in power grids, especially because of their huge capacity and long-distance transmission. The topography of a HVDC system is complicated compared to a HVAC/EHV system due to the presence of converting equipment. The range of the type of faults in a HVDC system is very wide. The faults may occur on the DC line, AC side of the rectifier, and AC side of the inverter due to which the identification of a particular fault condition becomes difficult and challenging. Soft computing techniques have shown smart performance in fault analysis with respect to speed and accuracy due to the popularity of tools like wavelet transform and S-transform.

Identification of DC faults in a multiterminal HVDC system has been proposed in [1] based on wavelet. The process of finding the fault location has not been discussed in this paper. A current protection scheme for a VSC–HVDC system has been proposed in [2,3] in which DFT has been used to obtain the harmonic content of the current signal. The fault location has not been determined in these papers. The extracted features have been used for training the neural networks. A method based on ST and SVM has been proposed in [4] for the identification of faults due to lightning in an UHVDC transmission system. A novel algorithm based on traveling-wave's natural frequency has been demonstrated in [5] where the natural frequencies of the transient current have been obtained to determine the fault location. The faults on the AC side and the effect of noise have not been considered here. Jian Liu et al. [6] present a new protection scheme for high-voltage direct-current (HVDC) transmission lines which only uses a specific frequency AC current (SFAC) at one of the line terminals during fault transients. The fluctuation characteristic of the root mean square (RMS) of SFAC is analyzed under different fault conditions by calculating a fluctuation coefficient using DFT. The exact location of the DC faults has not been obtained in this paper.

This chapter presents a technique of fault identification in a HVDC network and determination of DC fault location based on harmonic analysis. The proposed method involves discrete S-transform (DST) for feature extraction from the phase voltage and DC fault current signals. An expert system consisting of a combination of Probabilistic Neural Network (PNN) and Back Propagation Neural Network (BPNN) has been suggested for fault classification and estimating the fault location. All the simulations and the training have been conducted by programming in MATLAB.

The rest of the chapter is organized as follows. The simulation of the HVDC system and the different types of faults are described in Section 8.2. Section 8.3 provides the results of fault classification and determination of fault location by the combination of PNN and BPNN. Section 8.4 concludes this chapter.

8.2 SIMULATION OF THE HVDC SYSTEM AND FAULTS FOR THE STUDY

A six-pulse HVDC model has been simulated in MATLAB Simulink environment. A 500 MW (250 kV, 2 kA) DC interconnection is used to transmit power from a 315 kV, 5000 MVA AC network. The length of the DC transmission line is 300 km. The simulated model is shown in Figure 8.1. The sampling time of all the signals is taken to be 312.5 μs, and the time period of simulation in MATLAB is taken up to 0.5 sec. The sampling frequency is 3.2 kHz.

Ten types of faults (i.e., AG, BG, CG, AB, BC, CA, ABG, BCG, CAG, and LLL) have been simulated on the AC side of the DC transmission link at B1.

The faults have been initiated for different values of fault resistances starting from 0–100 Ω in steps of 20 Ω. DC line-ground (DCLG) fault has been simulated for different locations on the transmission line in steps of 10 km from the rectifier end. The range of variation of fault resistance is from 0–100 Ω in steps of 20 Ω. All the faults have been initiated at 0.2 sec and terminated at 0.3 sec. Two different fault inception times (T_F) have been considered in the simulation, i.e., 0.25 sec and 0.28 sec. Hence, the total number of AC fault simulations is 10×6×3 = 180 and that in the case of DC Faults is 29×6×3 = 522. The nature of the voltage profiles of different types of AC faults is shown in Figure 8.2. The waveform of the fault current during a DCLG fault is shown in Figure 8.3.

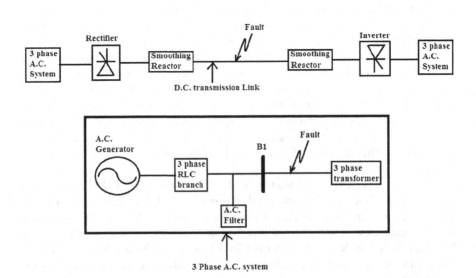

FIGURE 8.1   The simulated model of the HVDC system.

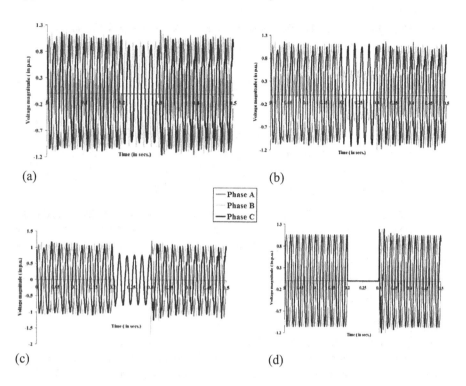

FIGURE 8.2 Voltage profiles for (a) AG, (b) AB, (c) ABG, and (d) LLL types of faults at B1 with $R_F = 0$ Ω and fault inception time = 0.2 sec.

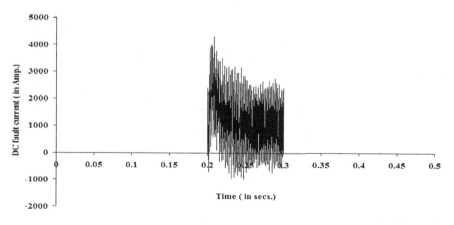

FIGURE 8.3 Fault current profile during a DCLG fault at 100 km from B1 with $R_F = 100$ Ω and fault inception time = 0.2 sec.

8.3 FAULT CLASSIFICATION AND DETERMINATION OF FAULT LOCATION

Fault identification has been achieved by a PNN. Two hidden layers have been used in the PNN. The first hidden layer is a radial basis transfer function, and the second one is a competitive transfer function. The size of the input layer for the PNN is 3×12, where 3 represents the input features (i.e., EA, EB, and EC of the three phases) and 12 is the number of fault types, including the no-fault condition. EA, EB, and EC are the signal energy of each phase on the rectifier end of the AC side of the DC link. EA, EB, and EC have been obtained in the same way as described in Chapter 6. The magnitudes of signal energy at a particular fault condition are summarized in Table 8.1

The normal signal and different faulty signals have been categorized as AG (1), BG (2), CG (3), AB (4), BC (5), CA (6), ABG (7), BCG(8), CAG (9), LLL (10), DCLG (11), and normal (12), respectively. The size of the output layer of the PNN is 1×12. The output of the PNN is one numeric integer ranging from one to 12 indicating the type of fault. The features of 12 voltage signals for each type of a.c. fault have been used for training the PNN. In the case of DCLG faults, 48 signals have been considered for training. The rest of the 534 data elements have been used for testing.

Once a DCLG fault is recognized, its location has been estimated by a combination of PNN and BPNN. The purpose of a new PNN is to determine the faulty section in the transmission line. BPNN estimates the location in the fault zone. It has been

TABLE 8.1
Magnitudes of Signal Energy for Different Types of Fault

	$R_F = 0\ \Omega, T_F = 0.2$ sec.			$R_F = 100\ \Omega, T_F = 0.2$ sec.		
	EA (p.u.)	EB (p.u.)	EC (p.u.)	EA (p.u.)	EB (p.u.)	EC (p.u.)
NORMAL	1.14	1.20	1.16	1.14	1.20	1.16
AG	1.39	1.31	1.20	1.18	1.22	1.20
BG	1.36	3.07	1.26	1.21	1.30	1.18
CG	1.30	1.36	3.71	1.16	1.27	1.23
AB	1.72	1.73	1.24	1.26	1.22	1.16
BC	1.25	2.95	3.02	1.16	1.34	1.20
CA	1.89	1.27	1.88	1.20	1.23	1.28
ABG	1.51	3.74	1.31	1.21	1.24	1.17
BCG	1.19	3.11	4.07	1.16	1.27	1.22
CAG	1.35	1.29	3.33	1.18	1.22	1.25
ABC	1.35	3.31	3.46	1.19	1.27	1.23
DC LG fault(10 km)	1.56	1.55	1.67	1.32	1.24	1.27
DC LG fault(50 km)	1.60	1.29	1.30	1.60	1.29	1.30
DC LG fault(250 km)	1.27	1.30	1.30	1.27	1.29	1.25
DC LG fault(290 km)	1.64	1.32	1.39	1.56	1.32	1.27

found from the simulations that the DC fault current consists of dominant harmonic components which follow a particular pattern in different sections of the transmission line. The architecture of the new PNN is the same as that has been used for fault classification. The two dominant frequency components of seven fault current signals corresponding to each section have been used as the input vector of the PNN. The size of the input layer is 2×3, where 2 represent the dominant frequency components and 3 is the number of faulty sections of the transmission line. The HVDC line has been segmented into three sections and categorized as Section-1 (1), Section-2 (2), and Section-3 (3). The size of the output layer of the PNN is 1×3 and the output is an integer ranging from one to three indicating the line section in which the fault has occurred. The harmonic components of 18 fault current signals out of 522 simulations have been used for training this PNN. The rest of the data have been used for testing. Once the faulty line section is identified, the distance of fault occurrence is estimated by BPNN. The frequency components of five current signals corresponding to five different fault locations in the affected section have been taken as inputs for training this BPNN. The output of the network is a numeric value. Three such BPNNs have been simulated in MATLAB corresponding to three sections of the HVDC line.

The results of fault classification and fault location estimation are summarized in Tables 8.2 and 8.3, respectively.

It is noticed in Table 8.2 that average accuracy of classification in the present study is 100%. The maximum error in the estimation of DCLG fault location obtained is 4.76%.

TABLE 8.2
Classification Results from PNN

Type of Fault	No. of Events	PNN Output		% Correct
		No. of Correct Predictions	No. of Wrong Predictions	
AG	6	6	0	100%
BG	6	6	0	100%
CG	6	6	0	100%
AB	6	6	0	100%
BC	6	6	0	100%
CA	6	6	0	100%
ABG	6	6	0	100%
BCG	6	6	0	100%
CAG	6	6	0	100%
ABC	6	6	0	100%
DCLG	474	474	0	100%

TABLE 8.3
Estimation of Fault Location from BPNN

Fault Resistance = 0 Ω			Fault Resistance = 20 Ω			Fault Resistance = 40 Ω		
Fault Location ((km)		% Error	Fault Location (km)		% Error	Fault Location (km)		% Error
Target Value	BPNN Output		Target Value	BPNN Output		Target Value	BPNN Output	
15	15.57	3.77	20	20.27	1.34	20	19.99	0.05
25	24.58	−1.68	40	41.74	4.35	40	40.00	0.00
35	34.22	−2.22	60	62.25	3.75	60	61.90	−3.17
45	45.35	0.79	80	80.00	0.00	80	80.68	−0.85
55	54.45	−1.00	100	100.00	0.00	100	100.00	0.00
65	66.84	2.82	110	110.00	0.00	110	110.00	0.00
75	75.89	1.18	120	119.99	−0.01	120	119.99	0.01
85	84.85	−0.17	130	129.38	−0.48	130	128.70	1.00
95	94.87	−0.14	140	140.11	0.08	140	140.18	−0.13
175	179.70	2.68	150	150.11	0.07	150	150.00	0.00
185	185.10	0.06	160	163.01	1.88	160	161.79	−1.12
195	190.04	−2.54	170	170.00	0.00	170	170.00	0.00
205	205.20	0.10	180	183.24	1.80	180	185.95	−3.31
215	217.14	1.00	190	190.00	0.00	190	190.00	0.00
225	230.00	2.22	200	200.00	0.00	200	200.00	0.00
235	234.17	−0.35	210	210.00	0.00	210	210.00	0.00
245	244.39	−0.25	220	230.48	4.76	220	230.42	−4.74
255	251.74	−1.28	230	239.70	4.22	230	239.65	−4.20
265	263.37	−0.62	240	241.80	0.75	240	241.18	−0.49
275	273.44	−0.57	280	288.00	2.86	280	288.00	−2.86
285	287.97	1.04	290	287.99	−0.69	290	288.99	0.35

8.3.1 EFFECT OF NOISE IN FAULT ANALYSIS

The PNN classifier and the BPNN have also been investigated by impregnating the entire testing data consisting of 570 elements with 20 dB white Gaussian noise. The simulated voltage waveforms and the fault current profile with noise are shown in Figure 8.4 and Figure 8.5.

The faults have been classified with 100% accuracy in the presence of noise. Table 8.4 shows the results of BPNN after considering 20 dB white noise in the simulated signals. The DCLG faults have been located with a maximum error of 4.01%.

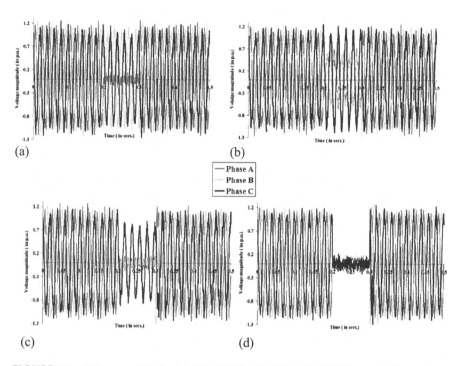

(a)   (b)

—Phase A
Phase B
—Phase C

(c)   (d)

FIGURE 8.4 Voltage profiles for (a) AG, (b) AB, (c) ABG, and (d) LLL types of faults at B1 with noise = 20 dB, R_F = 0 Ω, and fault inception time = 0.2 sec.

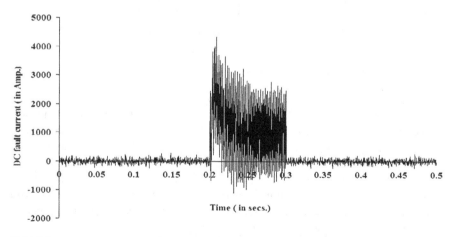

FIGURE 8.5 Fault current profile during a DCLG fault at 100 km from B1 with noise = 20 dB, R_F = 100 Ω, and fault inception time = 0.2 sec.

TABLE 8.4

Estimation of Fault Location from BPNN with 20 dB Noise

Fault Resistance = 0 Ω		% Error	Fault Resistance = 20 Ω		% Error	Fault Resistance = 40 Ω		% Error
Fault Location ((km)			Fault Location (km)			Fault Location (km)		
Target Value	BPNN Output		Target Value	BPNN Output		Target Value	BPNN Output	
15	15.57	3.77	20	20.02	0.09	20	19.93	0.35
25	24.13	−3.48	40	38.62	−3.46	40	40.10	−0.26
35	34.22	−2.22	60	60.01	0.02	60	58.98	1.70
45	44.06	−2.09	80	79.94	−0.08	80	80.69	−0.86
55	54.54	−0.83	100	97.64	−2.36	100	100.00	0.00
65	66.84	2.82	110	110.00	0.00	110	110.00	0.00
75	75.89	1.18	120	120.00	0.00	120	119.51	0.40
85	82.38	−3.08	130	127.43	−1.97	130	130.00	0.00
95	93.99	−1.07	140	139.78	−0.16	140	138.14	1.33
175	179.70	2.68	150	148.44	−1.04	150	150.00	0.00
185	185.10	0.06	160	157.68	−1.45	160	157.99	1.26
195	192.23	−1.42	170	170.00	0.00	170	170.00	0.00
205	205.20	0.10	180	177.38	−1.45	180	181.84	−1.02
215	211.13	−1.80	190	189.34	−0.35	190	189.97	0.01
225	230.00	2.22	200	196.01	−2.00	200	196.06	1.97
235	234.17	−0.35	210	210.00	0.00	210	210.00	0.00
245	244.54	−0.19	220	221.33	0.60	220	228.83	−4.01
255	256.57	0.62	230	235.00	2.17	230	238.80	−3.83
265	263.37	−0.61	240	240.12	0.05	240	240.04	−0.01
275	273.45	−0.57	280	284.50	1.61	280	283.50	−1.25
285	287.97	1.04	290	287.85	−0.74	290	287.95	0.71

8.4 CONCLUSION

The proposed method has produced satisfactory results in distinguishing between the AC and DC faults in a HVDC system. The location of DCLG faults has been determined with a maximum error of 4.76% without noise and 4.01% with 20 dB noise in the simulated signals. Only the voltage signals of the three phases on one terminal of the AC side have been used for fault classification. To obtain the location of the DCLG faults, only the harmonic components of the DC line current are required. DC fault locations have been determined from the combination of PNN and BPNN with acceptable accuracy. The effect of variation of fault resistance and fault inception time has been investigated in fault identification.

The suggested technique needs to be further implemented on a multiterminal HVDC system. The influence of variation of power angle, commutation failure in the rectifier, and inverter circuits needs to be incorporated during the study of fault analysis in the HVDC system.

8.A   APPENDIX

Generator: 315 kV, 5000 MVA, source inductance = 46.67 mH.
Transformer: 600 MVA, 315/210 kV
DC Transmission Line: R = 0.015 Ω/km, L = 0.792 mH/km, C = 14.4 nF/km

REFERENCES

1. K. De Kerf, K. Srivastava, M. Reza, D. Bekaert, S. Cole, D. Van Hertem, and R. Belmans, "Wavelet-based protection strategy for DC faults in multi-terminal VSC HVDC systems", *IET Generation, Transmission & Distribution*, vol. 5, no. 4, pp. 496–503, 2011. doi: 10.1049/iet-gtd.2010.0587.
2. Zheng Xiao-Dong, Tai Neng-Ling, J.S. Thorp, and Yang Guang-Liang, "A transient harmonic current protection scheme for hvdc transmission line," *IEEE Transactions on Power Delivery*, vol. 27, no. 4, pp. 2278–2285. doi: 10.1109/TPWRD.2012.2201509.
3. Xiaodong Zheng, Nengling Tai, Zhongyu Wu, and J. Thorp, "Harmonic current protection scheme for voltage source converter-based high-voltage direct current transmission system", *IET Generation, Transmission & Distribution*, vol. 8, no. 9 pp. 1509–1515, 2014, doi: 10.1049/iet-gtd.2013.0377.
4. H. Qin and F. Huang, "The identification of lightning faults in ±800kV UHVDC transmission lines using S-transform and SVM", *Proceedings of the 32nd Chinese Control Conference*, Xi'an, China, 2013, pp. 3859–3864.
5. Zheng-You He, Kai Liao, Xiao-Peng Li, Sheng Lin, Jian-wei Yang, and Rui-kun Mai, "Natural frequency-based line fault location in HVDC lines", *IEEE Transactions on Power Delivery*, vol. 29, no. 2, pp. 851–859, 2014, doi: 10.1109/TPWRD.2013.2269769.
6. Jian Liu, Nengling Tai, Chunju Fan, and Wentao Huang, "Protection scheme for high-voltage direct-current transmission lines based on transient AC current", *Generation, Transmission & Distribution, IET*, vol. 9, no. 16, p. 2633–2643, 2015. doi: 10.1049/iet-gtd.2015.0792.

9 Conclusion and Extension of Future Research Work

9.1 CONCLUSION

Signal analysis using soft computing techniques has evolved as a promising technique of fault analysis in power system networks. A plethora of research work has been conducted and myriad literature is still being published in this domain. It becomes difficult to design a neural network architecture with raw data as input. The features extracted from the raw data become more meaningful and specific which when used as the input vector of a neural network, the number of neurons in the middle layer reduces and the architecture becomes simple. In the present thesis, ST has been employed as a tool of feature extraction, and ANN has been implemented for fault classification and determination of fault location. To simulate the different power system models and to obtain the training data set, MATLAB-7 has been used along with the SimPowerSystems toolbox in Simulink. In order to train and analyze the performance of the neural networks, the Artificial Neural Networks Toolbox has been used extensively. The objective of this book is to reflect about the fundamental techniques of signal processing and their application in transient analysis of power system so that the readers can reapply the same or their modified versions in their own applications. The contributions of this book in the context of fault diagnosis are summarized as follows:

- The first chapter of this book after the Introductory one is on Power System Faults (Chapter 1). A brief description of the different types of faults and their analysis by conventional methods have been provided in this chapter. The advantages of application of signal processing techniques have also been highlighted in this chapter.
- Chapter 2 focuses on wavelet transform and its applications. Difference between stationary and non-stationary signals has been discussed with the help of examples. Simple examples of fault analysis by Wavelet Transform have been provided supplemented by MATLAB programs.
- Chapter 3 describes ST as a modification of STFT in which the width of the window function is inversely proportional to the frequency. The resolution of high-frequency components of a signal is improved in ST compared to STFT. ST can also be viewed as a phasor modification of WT in generating amplitude spectrum and phase spectrum independently. The output of ST is a complex matrix in which the row is the frequency and the column is the time. Unlike WT, ST is insensitive to noise. ST is widely applicable in heart sound analysis,

image watermarking, filter design, seismogram analysis, and power quality analysis. One of the major drawbacks of ST is its high computational burden due to generation of large amount of data. Computationally fast version of discrete ST is available to overcome this issue. The present research focuses on application of standard DST for fault analysis on overhead transmission lines. In the case of large amount of data, the series can be decomposed into smaller sections of data and DST can be applied in each section.

- Analysis of TFR of different time domain non-sinusoidal signals, both stationary and non-stationary, has been illustrated in Chapter 4 using ST, WT, and FFT. A. The magnitudes of harmonic amplitudes and their corresponding phase angles can be readily obtained from ST and as a result, an equation of a non-sinusoidal signal can be framed. Harmonic analysis of inrush current waveform of a saturated transformer using ST has been demonstrated in which an equation of the current signal has been derived from the information of the harmonic amplitudes and their phase angles. The resultant waveform has been plotted from the derived equation of the signal and its profile closely resembles with the simulated inrush current signal. A comparative analysis of ST, WT, and FFT has been summarized. Unlike FFT, both ST and WT would give TFR of any signal. ST and FFT are immune to noise. WT decomposes a signal into several frequency bands. ST has the highest computational burden.

- ANN is a widely accepted tool of fault identification, classification, and finding the location of the fault in a power system network. PNN is an implementation of statistical algorithm in which the operations are organized into multilayered feed forward network with four layers: input layer, pattern layer, summation layer, and output layer. A PNN is predominantly a classifier since it can map any input pattern to a number of classifications. The main advantages that discriminate PNN are: fast training process, an inherently parallel structure, guaranteed to converge to an optimal classifier as the size of the representative training set increases, and training samples can be added or removed without extensive retraining. The LM algorithm of BPNN has estimated the fault locations with acceptable accuracy and speed. In the present book, fault location has been determined from a BPNN. Chapter 5 describes the concept of BPNN and PNN in detail. The simulated PNNs in the present research have classified different faults in Chapters 6–8 with an average accuracy of 97.6%–99.6%. All the PNNs have been rigorously tested with noisy data, and the range of accuracy of classification is 96.8%–98.7%. The BPNN architectures simulated in Chapters 6–8 have estimated the fault locations with a maximum error in the range of 4.46%–4.75% without noisy signals and 4.13%–4.35% with noisy signals. The results indicate that in the present work the PNN and BPNN architectures have satisfactorily classified and located all the simulated faults.

- In Chapter 6, a method of fault identification and estimation of fault location on overhead transmission line has been proposed based on ST and ANN. The power system model comprises of a two-terminal single-circuit system having a sending end and a receiving end. Ten types of faults for different locations, fault resistances, and fault inception angles have been simulated for generating two types of data sets: testing data and training data. The current signal of each phase of both the terminals has been considered to extract features of fault classification. The number of features used for fault detection is six and for estimation of fault location the number is one. The PNN designed for this purpose has classified all the testing data with an average accuracy of 99.6%. The PNN has been further tested by incorporating white Gaussian noise of 20 dB in the testing data, and an average accuracy of 98.7% has been obtained. Once the faulty phase(s) has been identified the fault location has been obtained from a BPNN. The voltage signal of only the effected phase(s) of the sending end of the network has been considered to extract features for training the BPNN. The maximum error achieved for the entire testing data set is 4.46%. The same BPNN is further investigated with simulated noisy voltage signals and the maximum error obtained is 4.35%. The results obtained indicate that the proposed method is capable of detecting a fault and obtain its location on overhead transmission line with acceptable speed and accuracy.

- Signal analysis-based fault classification has been suggested in Chapter 7 in which the voltage signal of each phase at a single terminal of a network has been considered for feature extraction. The architecture of PNN remains the same as described in Chapter 6. Only three features are needed for training the PNN resulting in further reduction of memory requirement and computation time. A power system network with unbalanced loading has been simulated in which the faults have been classified with 99.6 % accuracy for pure signals and 96.8% accuracy for noisy signals. The method of finding fault location proposed in Chapter 6 has been implemented in this system. The maximum error achieved in fault distance estimation is 4.75% without noise and 4.27% for noisy signals. A IEEJ West 10 model has been further considered in the study to validate the proposed technique. The faults have been classified in the system with an average accuracy of 97.6%. As this is a multiterminal system, fault location estimation has been performed by a combination of PNN and BPNN architectures. The faults have been located with a maximum error of 4.13%. In this chapter, the proposed scheme has produced satisfactory results in both the systems.

- Chapters 6 and 7 have discussed the method of fault identification and obtaining its location in a.c. systems. The diagnosis of faults in a HVDC system is another challenging task. Chapter 8 presents signal analysis-based fault classification in a six-pulse HVDC system. All the ten types of a.c. faults (as described in Chapters 6 and 7) and DC Line-Ground (DCLG) have been simulated. The proposed PNN architecture has distinguished between the a.c. and d.c. faults with 100% accuracy. Once the DCLG faults have been identified, they have been located by a combination of PNN and BPNN. The harmonic components of DC line current have been used as the training features of the PNN and BPNN for obtaining fault location. The location of DCLG faults has been determined with a maximum error of 4.76% without noise and 4.01% with 20 dB noise in the simulated signals. The proposed scheme of fault identification and determination of fault location has shown satisfactory performance in the simulated HVDC system.

9.2   FUTURE WORK

The present book/monograph provides a wide opportunity for research work in Transient analysis of power system via soft-computational techniques. A few of the major possible extensions of the research work have been listed below:

- The results of fault analysis obtained in Chapters 6–Chapter 8 must be compared with other versions of ST and WT, such as Hyperbolic ST, Discrete Orthonormal ST (DOST), and Fast S-Transform (FST)
- The proposed technique in Chapter 6 should be tested in a multiterminal and multifeed network. The effect of transformers, CT, and PT needs to be studied by incorporating them in the present system. The results can be further investigated by using generalized ST and other fast versions of discrete ST.
- A modified IEEJ West 10 machine system has been considered in Chapter 7 which is a multiterminal, multifeed system. As the simulation of the actual IEEJ West 10 machine system model requires an advance version of MATLAB/ EMTP software the results obtained in Chapter 7 can be further validated on the actual model where all the interconnecting lines are double-circuit systems.
- In Chapter 8, a six-pulse HVDC system has been simulated in which a method of fault classification and estimation of DCLG fault location has been suggested. The proposed technique needs to be tested on a multiterminal, 12-pulse HVDC system. The effect of commutation failure of the rectifier and inverter circuits on the AC, side of the HVDC link needs to be investigated during the fault analysis.

Index

Page numbers in *italics* refer to illustrations and those in **bold** refer to tables.

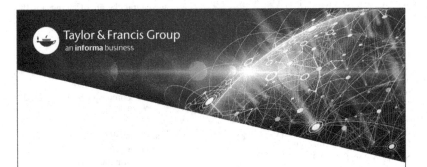

Printed in the United States
by Baker & Taylor Publisher Services